Anatomy 1
Laboratory Text

Human Anatomy

Santa Rosa Junior College

Manual prepared by: Robert Rubin, Susan Wilson, Danielle King, Pete Arnold, Nick Anast, Mike Henry, Caitlyn Beaton, Tony Graziani, Colette Bizal, and Mike Lindh-Cabrera

Published by Arbor Crest Publishing

Cover mage by Andreas Vesalius's *De humani corporis fabrica* (1543), page 163.

Table of Contents

Chapter 1

Introduction to Anatomy, Cells

ANATOMICAL TERMS OF REFERENCE, BODY CAVITIES, ORGAN SYSTEMS, and CELLS

For the student, the first laboratory in a course in human anatomy may be the most important in developing concepts and landmarks that will aid in your understanding of all successive laboratories. This laboratory will introduce you to some basic anatomical terminology which is used by anatomists to refer to various aspects of body structure. Such terms are necessary to enable anatomists and medical personnel to understand one another when aspects of body structure are being discussed. The second part of this laboratory will introduce you to body cavities and organ systems. You will also review the parts of the cell and their functions. A few days after this lab, take the self-quiz on p. 7 to assess how much you remember.

Note: helpful information can be found throughout the semester on **Danielle King's website:** https://sites.google.com/view/srjcanatomy1

I. Anatomical Terminology

A. Anatomical Position

All reference to anatomical structures in terms of their location, positioning, and orientation is made assuming the body is in a given position, i.e., the **anatomical position.** In the anatomical position, the subject is standing erect (upright position) and facing forward with the upper extremities (upper limbs) at the sides, the palms of the hands facing forward, and the feet flat on the floor with toes pointing forward.

B. Body Sections (or planes)

The structural plan of the human body may be described with respect to <u>planes</u> (imaginary flat surfaces) passing through it. Planes are frequently used to show the anatomical relationship of one or more structures in one region of the body relative to another.

Body sections commonly referred to are:

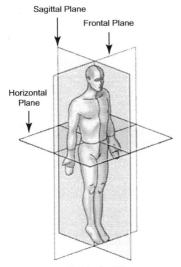

Figure 1–Body Planes

1. **Sagittal (median)**
   a. **Midsagittal** – A vertical plane that passes through the midline of the body and divides the body into *equal* right and left sides.
   b. **Parasagittal** – A vertical plane that does not pass through the midline of the body and which divides the body into *unequal* right and left sides.
2. **Frontal (coronal)** – A vertical plane at a right angle to the midsagittal plane which divides the body or an organ into anterior and posterior portions.
3. **Transverse (horizontal)** – A plane that is parallel to the ground (at a right angle to the midsagittal plane) and divides the body or an organ into superior and inferior portions. (This plane may be referred to as a "cross section.")
4. **Oblique** – slanted, at an angle

C. Terms Relating to the Relative Position of Body Parts

The terms used for humans (two–legged animals) differ slightly from those used for four–legged animals. A knowledge of the terms of location and position is necessary for referencing structures and in order to follow instructions in future laboratory or clinical work.

Note how the following terms form antonymous pairs (i.e., the members of the pair have opposite meanings).

1. **Superior** (cranial, cephalic) – more toward the head end of the body, or toward the upper end or surface of a body part.
   Inferior (caudal) – more toward the feet, or toward the lower end or surface of a body part.

2. **Anterior** (ventral) – more toward the front surface of the body or an organ.
   Posterior (dorsal) – more toward the back or rear surface of the body or an organ.

3. **Medial** – toward the midline of the body or an organ, or as viewed from the midline.
   Lateral – away from the midline of the body or an organ, or as viewed from the side.

4. **Proximal** – nearer the point of attachment of a limb to the body trunk; nearer to the point of origin.
   Distal – further from the point of attachment of a limb to the body trunk; further from the point of origin.

5. **Superficial** – more near the surface of the body.
   Deep – further away from the surface of the body (i.e., more internal).

   Be able to demonstrate on your body, and on models and charts in the lab, all of the bold–faced terms listed above.

II. Body Cavities

Spaces within the body that contain various internal organs are called **body cavities** (see Figure 2 on page 5). The principal body cavities are listed below:

A) Posterior (dorsal) body cavity
　　1. Cranial cavity
　　2. Vertebral (spinal) cavity
B) Anterior (ventral) body cavity (coelom)
　　1. Thoracic cavity
　　　　a. Right pleural cavity
　　　　b. Left pleural cavity
　　　　c. Mediastinum
　　　　　　1) Pericardial cavity
　　2. Abdominopelvic cavity
　　　　a. Abdominal cavity
　　　　b. Pelvic cavity

Mediastinum: the space between the pleural cavities, extending from the sternum (breast bone) to the backbone and from the thoracic inlet to the diaphragm is called the **mediastinum.** It contains all structures in the thoracic cavity, except the lungs themselves and those portions of the pleural sacs which cover the lungs and line the inner surface of the chest wall.

Structures included in the mediastinum are:
heart, thymus gland, esophagus, trachea, many large blood and lymphatic vessels

Using the torso model or chart locate the muscular organ of respiration, the **diaphragm**. This sheet of muscle divides the ventral cavity into the thoracic cavity (superior) and the abdominopelvic cavity (inferior). Identify the other cavities on the torso model.

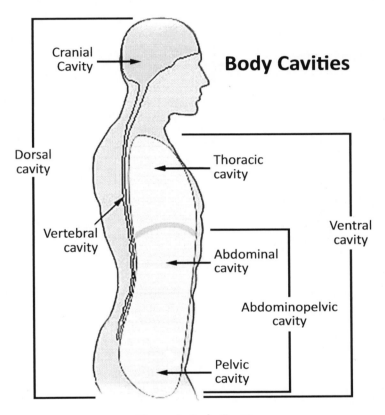

Cranial Cavity

Body Cavities

Dorsal cavity

Thoracic cavity

Ventral cavity

Vertebral cavity

Abdominal cavity

Abdominopelvic cavity

Pelvic cavity

Figure 2–Body Cavities

III. Systems of the Human Body

We are studying the human body using a systemic approach. The course explores each organ system (organ system = a collection of organs organized together to carry out a common function or set of functions i.e., **nervous system** = brain, spinal cord, nerves, ganglia). Each body system has specific functions and usually consists of several organs. In turn systems work together to regulate and coordinate body–wide function.

Use your textbook to identify the major body systems and their organs on models and charts. Complete the chart on the next page. This will help you learn to identify some organs, the systems they belong to, and the body cavities in which they are found.

Organ	Most Specific Space Organ is Contained Within	Organ System to which this Organ Belongs	Two Other Organs in the Same System	Two Functions of this System
1 Esophagus				
2 Trachea				
3 Heart				
4 Stomach				
5 Uterus				
6 Colon				
7 Liver				
8 Kidneys				
9 Urinary Bladder				
10 Spleen				
11 Adrenal Glands				
12 Pancreas				
13 Lungs				
14 Brain				

Questions for Review

1. Match the terms on the right with the definitions on the left. You may use one answer more than once or not at all.

( ) A cut that divides body into right and left sides

( ) The belly or front of torso

( ) Cavity containing brain

( ) Cut dividing body into anterior and posterior

( ) Space between membranes that surround the heart

( ) Cavity containing the heart and lungs

( ) Cut dividing body into dorsal and ventral parts

( ) Away from or to the side of the midline

( ) Space between lungs

a. Anterior
b. Posterior
c. Lateral
d. Pericardial cavity
e. Cranial cavity
f. Mediastinum
g. Thoracic cavity
h. Sagittal section
i. Frontal or coronal
j. Horizontal or Transverse

2. Name the most specific space in which the organs listed below are found.

a. Brain _____

b. Stomach _____

c. Heart _____

d. Small Intestines _____

e. Spinal cord _____

f. Pancreas _____

g. Lungs _____

h. Esophagus _____

i. Uterus _____

j. Liver _____

3. Fill in the correct directional term for location and position. (There may be more than one correct answer.)

a. The human head is _____ to the neck.

b. The elbow is _____ to the wrist.

c. The neck is _____ to the head.

d. The eye is _____ to the nose.

e. The mouth is _____ to the ears.

IV. Cells and Organelles

The cell is the primary and fundamental functional and structural unit of all living organisms. This basis of all structure and function is the platform from which an understanding of tissues, organs, and organ systems is developed. All of the activities of body systems and individual organs can be traced to cellular function.

The cells of the human organism are highly diverse and their differences in shape, size, and internal composition reflect their specific roles in the body. While cells differ in some respects, they do share several basic structures and functions. For example, all cells have the ability to metabolize, reproduce, grow, and to respond to stimuli.

Because most cells are small in size, not all of their features are visible under the light microscope. However, four areas or structures common to most cells can be distinguished using this instrument: The **nucleus**, the **plasma membrane**, the **nucleolus** and the **cytoplasm**. The nucleus is usually seen as a single round or oval structure near the center of the cell. It contains one or more smaller, darker bodies known as **nucleoli**. The nucleus is surrounded by intracellular material known as the cytoplasm and the entire mass is enclosed by the plasma membrane. Smaller cell structures known as **organelles** are found in the cytoplasm but are too small to be visible under the microscope. Intracellular fluid within the cytoplasm is known as **cytosol**.

Basic Cell Structure

The **plasma membrane**, **cytoplasm**, **nucleus**, and **nucleolus** are visible in the light microscope. Learn to identify these structures using prepared slides and the cell model. The remaining structures (#5 – 9) should be identified only on the cell model, as they cannot be identified with a light microscope.

1. Plasma membrane
2. Cytoplasm
3. Nucleus
4. Nucleolus

5. Ribosomes
6. Rough endoplasmic reticulum
7. Smooth endoplasmic reticulum
8. Golgi complex
9. Mitochondrion

CHAPTER 2

Microscopes, Major Tissues, Epithelial Tissue

Microscopes, Major Tissues, Epithelial Tissue

I. The Microscope

Gross anatomy is that branch of anatomy which deals with structures which can be observed by the unaided eye. When you work with the cadaver and all of its visible components, you are studying gross anatomy. Histology, by comparison, is the study of the detailed anatomy of tissues at the level of the cell. Because cells are exceedingly small, they require a microscope to see them. The microscopic study of tissues is called **histology** (histo = tissue).

Your success in performing laboratory work will depend in part upon the physical equipment at your disposal and your capacity to use it intelligently. The most valuable and complex instrument you will use is the microscope. If you learn to use it well and to care for it properly, it will be a source of great help to you. Because your microscope is a costly instrument and will be used by other students, you are obligated to leave it in good condition and in its proper place.

Study the list of microscope components and their functions **before** using the microscope.

Instructions for Microscope Use

1. Get your microscope from the cabinet, carrying it with one hand on the arm and the other under the base.

2. Place the microscope on the table with the arm toward you. Gently pick it up and turn it around so that the oculars face you, **without** sliding the microscope across the table.

3. Clean the lenses of oculars (eyepiece) and objectives with lens paper. The rubber rim of the oculars should be folded down if you wear glasses, extended if not.

4. Place the 4x objective in place under the body tube.

5. Adjust the light until you see a clear field of light when you look into the oculars. Change the interpupillary distance by gently moving the base of the oculars closer together or farther apart, just like you would with binoculars.

6. Always keep both eyes open when looking through the microscope. This practice helps to avoid eyestrain, and ensures that you can see the pointer(s) used during exams.

7. Place the slide to be studied on the stage, being sure the coverslip is up. Use the stage control knobs to move the slide so that the material to be observed is directly under the scanning objective (4x).

8. Move the stage up until it is nearly to its highest position. Now look through the ocular and, using the coarse adjustment knob, slowly lower the stage until the material comes into focus.

9. You may need to readjust the light for best definition. Light intensity can be varied with the rheostat switch of the in–base illuminator, the iris diaphragm, and/or the condenser. Initially open the iris diaphragm to allow about 75% of the light to enter and raise the condenser almost all the way up. Once these settings are in place, change light intensity with the rheostat to reduce glare and improve contrast.

10. If you wish to see the specimen at greater magnification, use the low power objectinve (10X). Start by centering in the field that part of the material which you wish to observe; then carefully turn the low power (10X) objective into position. Readjust your lighting and focus with the fine adjustment if necessary.

11. Use the diopter adjustment located on the left ocular to compensate for the difference in focusing ability between your two eyes.

12. It is important to follow the above procedure for efficient use of the microscope. Do not try to use high power without first getting the material into focus under lower power.

13. Do not use the coarse focus knob when using high power (400x and 630x). Many slides and lenses are damaged in this way.

14. When returning the microscope to the case, be sure to remove the slide, turn off the light and have the scanning objective in place under the body.

15. Calculation of the amount of the magnification of an object is performed by multiplying the power of the ocular (10x) by the power of the objective in use. For example, the use of a 40x objective would yield a magnification of 400x (40x • 10x = 400x).

Microscope Parts

Following the directions given in lecture and above, familiarize yourself with the microscope and its proper use. Understand the method for calculating magnification. Learn the following parts of the microscope and their functions:

1. In–base illuminator
   a. Rheostat
2. Eyepiece (ocular)
3. Objective Lenses:
   a. Scanning objective (4x)
   b. Low power objective (10x)
   c. High power objectives (40x and 63x)
4. Mechanical Stage
   a. Stage control knobs
5. Condenser
   a. Iris diaphragm and control ring
   b. Condenser adjustment knob
6. Adjustment knobs
   a. Coarse focus knob
   b. Fine focus knob

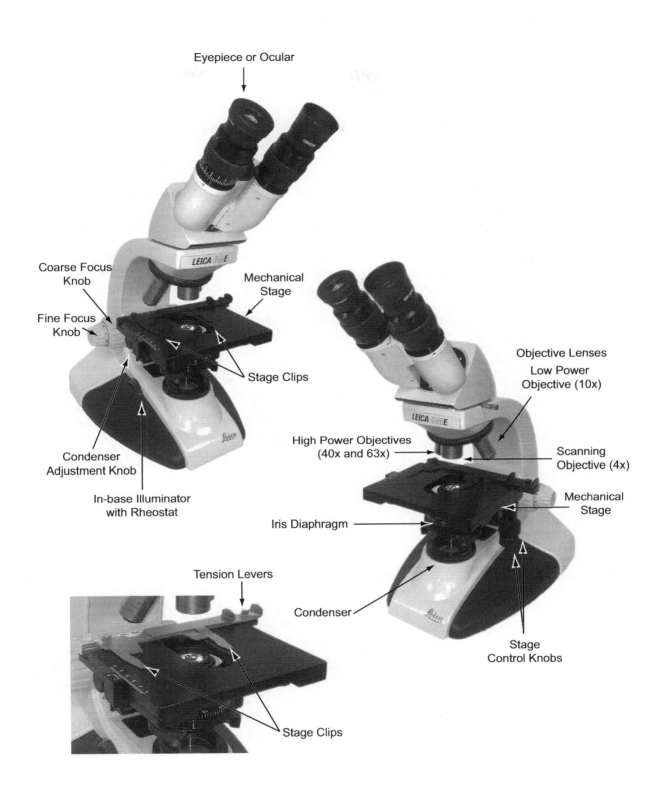

Eyepiece or Ocular

Coarse Focus
Knob

Fine Focus
Knob

Mechanical
Stage

Stage Clips

Objective Lenses
Low Power
Objective (10x)

High Power Objectives
(40x and 63x)

Scanning
Objective (4x)

Condenser
Adjustment Knob

Mechanical
Stage

In-base Illuminator
with Rheostat

Iris Diaphragm

Tension Levers

Condenser

Stage
Control Knobs

Stage Clips

THE MICROSCOPE

Microscope Components and their Functions *		
Microscope Component	**Location or Description**	**Functions**
Ocular (eyepiece)	Uppermost series of lenses	Magnification
Nosepiece	Revolving assembly	Holds objectives
Scanning power objective	Shortest objective, magnifies 4x	Magnification
Other objectives	Microscope also has 10x, 40x & 63x objectives	Magnification
Condenser	Lens system located below central stage opening	Concentrates and directs light beam through specimen
Stage	Platform upon which specimens for examinations are placed	Specimen support
Mechanical stage	Control device that permits moving slides left to right, forward & backward on stage	Manipulation of specimen's location
Iris diaphragm	Located beneath stage in association with condenser unit, controlled with a lever	Regulates brightness or intensity of light passing through lenses
Coarse focus adjustment knob	Large knob on arm, below stage	Used for preliminary and coarse focusing by raising or lowering stage
Fine focus adjustment knob	Smaller knob, below stage	Used for final or fine focusing by raising or lowering stage
Condenser adjustment knob	Control knob located below stage	Used to obtain full illumination by raising or lowering condenser
Base	Heavy, bottom portion on which the instrument rests	Microscope support
In base illuminator	Light switch, rheostat	Turn on, adjust light intensity
Arm	Vertical, curved portion of microscope, used in carrying instrument	Microscope support

* This list is for reference purposes, the list of microscope parts to learn for exams is on the previous page

II. Major Tissue Types

Tissues are defined as aggregates of cells which are similar in structure, and work together to perform a specific activity. The human body is composed of four major kinds of tissues:

A. **epithelial tissues**
B. **connective tissues**
C. **muscular tissues**
D. **nervous tissues**

Each type performs a specialized function.

Epithelial tissues occur as sheets of cells and form the linings or coverings of the inner and outer surfaces of organs (the membranous epithelia). They also comprise the major tissue of most glands (glandular epithelium). Epithelial tissues are characterized by their cells being closely packed together with little or no intercellular substance between them.

Connective tissues form the supportive tissues of the body. Common examples are bones, cartilage, and tendons. They also form the fibrous networks which hold the various organs together and prevent other tissues from tearing or losing structural integrity.

Muscle tissues are the contractile tissues. Muscle cells have the unique ability to dramatically shorten their length in response to nerve impulses, and when attached to body parts, are able to move them.

Nervous tissues are composed of cells which carry electro-chemical signals. They form a rapid communication network throughout the body.

The four tissue types will be studied as follows:

Epithelial –	Chapter 2
Connective	
Proper and Cartilage–	Chapter 3
Bone –	Chapter 5
Blood –	Chapter 7
Muscle –	Chapter 6
Nervous –	Chapter 8

It is possible to observe all four tissue types in many organs. This will be done when we examine the Integumentary System in Chapter 4.

III. Epithelial Tissue

Study each of the following examples of epithelial tissue types with your microscope. You will need to refer to your text and to an atlas of histology for help as you go along. For each tissue type listed below, you should be able to:

- identify the specific tissue type
- give two examples of where it occurs in the body
- state the typical function(s) of that tissue

A. Covering and Lining Epithelium

1. **Simple squamous epithelium** – Observe lining *blood vessels* (especially arteries; see artery, vein & nerve slide) and lining the *glomerular capsule* (see kidney slide)

2. **Simple cuboidal epithelium** – Observe in *kidney nephrons*

3. **Simple columnar epithelium** – Observe lining *fundic stomach*

4. **Simple columnar epithelium with goblet cells and microvilli** – Observe lining *jejunum* (a region of small intestine)

5. **Pseudostratified ciliated columnar epithelium with goblet cells** – Observe lining *trachea*

6. **Stratified squamous epithelium** – Observe lining *esophagus* (note that this is a non-keratinized tissue; we will observe keratinized tissue in the epidermis of the skin)

7. **Transitional epithelium** – Observe lining *ureter*

B. Glandular epithelium

1. Exocrine glandular epithelium – Observe in *submandibular/submaxillary salivary glands* (note both the secretory units and the **ducts**, which are formed by a covering and lining epithelium)

2. Endocrine glandular epithelium – Not observed during this lab.

CHAPTER 3

Connective Tissue

CONNECTIVE TISSUE

Connective Tissue is one of the four major tissue types found in the body. It is characterized by having a large amount of extracellular matrix in which **cells** are embedded. The matrix consists of **fibers**, which are secreted by the cells, and **ground substance.** The nature of the ground substance determines the categories of connective tissue: solid (bone), gelatinous (connective tissue proper and cartilage), and fluid (blood). Connective tissues serve to connect, support and protect other tissues and organs.

In this lab we are examining six specific types of connective tissue proper, three specific types of cartilage, and one specific type of bone tissue. For each one be able to recognize the tissue, identify its location in the body, and know its function(s).

A. Loose Connective Tissue Proper
1. **Areolar Connective Tissue**
   a. Fibroblasts
   b. Mast cells
   c. Plasma cells (see demonstration slide)
   d. Elastic Fibers
   e. Collagen fibers

2. **Adipose Connective Tissue** – Observe in *hypodermis* (see scalp/thin skin slide; note that it's mixed with areolar CT in this location)
   a. Adipocytes
   b. Collagen fibers

3. **Reticular Connective Tissue** – Observe in stroma of *lymph node*
   a. Reticular fibers (*only visible with special stain*)

B. Dense Connective Tissue Proper
1. **Dense Regular Connective Tissue** – Observe in *tendon, ligament*
   a. Fibroblasts
   b. Collagen fibers

2. **Dense Irregular Connective Tissue** – Observe in *dermis* (see scalp/thin skin slide)
   a. Collagen fibers

3. **Elastic Connective Tissue** – Observe in wall of *aorta* (mixed with smooth muscle tissue)
   a. Elastic fibers

C. Cartilage
1. **Hyaline Cartilage**
2. **Elastic Cartilage**
3. **Fibrocartilage**

For all of the above cartilage tissues note:
   a. Chondrocytes
   b. Lacunae
   c. Matrix
      1) Ground substance
      2) Fibers: collagen or elastic
   d. Perichondrium: surrounds pieces of hyaline and elastic cartilage only
      1) Fibroblasts
      2) Chondroblasts

D. Bone
1. **Compact Bone Tissue** – histological slides
   a. Osteons (Haversian systems)
      1) Concentric lamellae
   b. Interstitial lamellae
   c. Lacunae: house osteocytes in life
   d. Central canals (Haversian canals): house AVLN in life
   e. Canaliculi: house projections of osteocytes in life
   f. Perforating canals (Volkmann's canals): house AVLN in life

2. **Bone** – models
   Compare the models with the slides of bone. Note the structures listed above, and in addition:

   a. Periosteum
   b. Endosteum
   c. Sharpey's fibers (perforating fibers)
   d. Circumferential lamellae
   e. Compact bone tissue
   f. Spongy bone tissue
      1) Trabeculae

CHAPTER 4

Integumentary System

The Integumentary System

I. Microscopic Anatomy (Histology) of Skin

Examine the models and chart of the skin and become familiar with the orientation of the integument and its various regions.

Examine slides of **thick skin** from the palm or sole and **thin skin** from both the scalp and the axilla (underarm). Identify the following:

A. Epidermis
1. Stratum basale
2. Stratum spinosum
3. Stratum granulosum
4. Stratum lucidum (thick skin only)
5. Stratum corneum
6. Free nerve endings (chart only)
7. Friction ridge (model only)

B. Dermis
1. Papillary layer
   a. Dermal papilla
   b. Meissner's corpuscles (tactile corpuscles)
   c. Free nerve endings (chart only)

2. Reticular layer
   a. Pacinian corpuscles (lamellated corpuscles))
   b. Ruffini corpuscles (chart only)
   c. Free nerve endings (chart only)
   d. Sebaceous glands
   e. Merocrine (eccrine) sweat glands
   f. Apocrine sweat glands (seen only on axillary skin slide)
3. Hair
   a. Root - below surface of skin
   b. Shaft - above surface of skin
   c. Follicle
   d. Bulb of the follicle
   e. Arrector pili miscle - made of smooth muscle tissue
C. Hypodermis (subcutaneous layer, superficial fascia)
   1. Adipose connective tissue (mixed with areolar CT)
   2. Pacinian corpuscles (lamellated corpuscles)

Two types of sweat glands are found in the body. **Merocrine (eccrine)** sweat glands are the type most widely distributed in humans. They are simple tubular glands that are found everywhere except genitalia and lips. They are located in the deep dermis and their ducts terminate in pores at the surface of the epidermis. These glands primarily secrete watery sweat which evaporates and removes body heat from the skin's surface. **Apocrine** sweat glands are limited in their location to the pubic and anal regions as well as the axillae and the pigmented region of the breast. These glands begin to function at puberty and their secretions pass through ducts that empty into hair follicles.

HELPFUL HINTS: FYI

Many anatomical organs are constructed as tubes with cellular/tissue walls and a hollow center called the lumen. Some organs of this design are large, macroscopic structures such as the trachea, intestines and uterus, while others are small, often microscopic, and course within the connective tissues of larger organs. In this latter group are blood vessels (veins, arteries and capillaries) and the ducts of exocrine glands such as the sweat glands and sebaceous glands found in the skin.

Two problems often encountered by students when viewing microscopic tubular organs for the first time through the microscopic are: 1) Three-dimensional interpretation: How do you get an idea of the 3-D structure of an object when you see it in 2-D, and how has it been cut/sectioned to expose this view? 2) Specific identification: How do you tell different tubular organs apart that are structurally similar?

Three-dimensional interpretation

Because tubular organs course through tissues like pieces of wet, folded, flexible spaghetti rather than like dry, straight, rigid pieces, any number of cuts, planes, and views are possible when a slice of tissue is placed on a microscopic slide. Realize that any microscopic section will contain only a small bit of the entire tube. Note the diagrams below that illustrate and explain the possible sections and planes of cut that the tube may present to you:

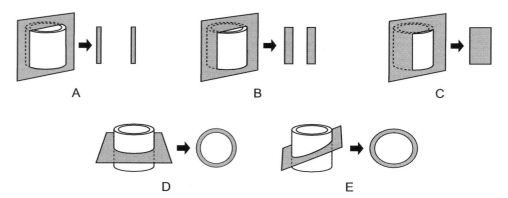

Diagrams showing the appearance of sections of a straight tube cut in various planes.
A, B, and C are longitudinal sections. D is a transverse or cross section.
E is an oblique section.

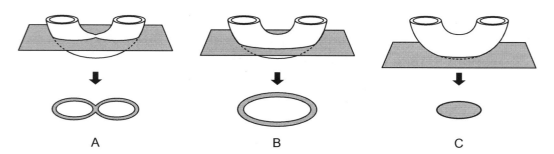

Diagrams showing the appearance of sections of a curved tube cut in various planes relative to the lumen. A and B include the lumen; C does not.

CHAPTER 5

Skeletal System

SKELETAL SYSTEM

Gross Anatomy of the Human Skeleton

A. Organization of the Skeleton

The adult human skeleton, containing 206 bones, may be subdivided as follows:

1. **Axial Skeleton**

   Skull (will be studied with Nervous System)
   Auditory ossicles (will be studied with Special Senses)
   Hyoid
   Vertebral column
   Thorax
   > Sternum
   > Ribs

2. **Appendicular Skeleton**

   Pectoral (shoulder) girdle
   Bones of the upper limbs
   Pelvic (hip) girdle
   Bones of the lower limbs

B. Bone – Gross Anatomy of a Long Bone

Obtain a femur that has been sectioned along its longitudinal axis. Note the following generic structures, which can be found on many bones:

1) Diaphysis (shaft)
2) Proximal epiphysis
3) Distal epiphysis
4) Compact bone tissue
5) Spongy bone tissue
   a. Trabeculae
6) Marrow (medullary) cavity
7) Articular surfaces
8) Points of ligament and tendon attachment (surface features)
9) Epiphyseal line
10) Nutrient foramina

Learning Bones and Surface Features

TAKE CARE when handling skulls, bone specimens, and models. PLEASE use only wooden probes provided as pointers, and use those gently, to protect against damage to delicate bone. The use of pens and pencils in the past has damaged bones and models.

Be able to:

- identify all structures on the following lists (pages 29–33). If a feature can be found on more than one bone, the name of the bone must also be included when one is asked to identify that feature on a lab exam.
- distinguish left versus right for ribs and all appendicular bones
- give the function for all surface features.
    - One common function is muscle attachment.
        - You do not need to know the muscle names until we study the Muscular System in lab; however, it would be helpful to reference page 40.
        - It is also helpful to look at the red and blue painted surfaces of the skeletons to visualize where muscles attach.
    - If the function is articulation, know specifically what that feature is articulating with.
    - If the function is to house another structure (such as a blood vessel or nerve), know specifically what is housed by that feature.

C. Bones of the Human Skeleton

1. AXIAL SKELETON

a. Hyoid

b. **Skull**

The skull is divided into cranial and facial bones. The cranial bones are those that surround the cranial cavity; all others are facial bones.

Spend about half an hour locating the individual bones before moving on.

1) Cranial bones
   a) Single
      (1) Frontal
      (2) Ethmoid
      (3) Occipital
      (4) Sphenoid
   b) Paired
      (1) Parietals
      (2) Temporals

2) Facial bones
   a) Single
      (1) Mandible
      (2) Vomer
   b) Paired
      (1) Nasals
      (2) Lacrimals
      (3) Maxillae
      (4) Zygomatics
      (5) Palatines
      (6) Inferior nasal conchae (turbinates)

Next, located the following features on a complete skull.

3) Sutures
   a) Sagittal suture
   b) Coronal suture
   c) Lambdoidal suture
   d) Squamous suture

4) Fontanels – these will be visible on the fetal skulls only
   a) Anterior (frontal) fontanel
   b) Posterior (occipital) fontanel
   c) Anterolateral (sphenoid) fontanels
   d) Posterolateral (mastoid) fontanels

5) Cavities
  a) Cranial cavity
  b) Nasal cavity

6) Paranasal Sinuses
  a) Frontal sinuses
  b) Maxillary sinuses
  c) Sphenoidal sinus
  d) Ethmoidal air cells

7) Foramina
  a) Foramen magnum
  b) Carotid canal
  c) Jugular foramen
  d) Optic canal

Finally, learn the specific surface features for each individual bone.

8) Surface features of individual bones
  a) Frontal
    (1) Orbital plate (orbital surface)
    (2) Frontal sinuses
    (3) Zygomatic process of the frontal
  b) Parietal
    (1) Groove for superior sagittal sinus *
  c) Occipital
    (1) Foramen magnum
    (2) Occipital condyles
    (3) Jugular foramen (between occipital and temporal bone)
    (4) Groove for superior sagittal sinus*
    (5) Grooves for transverse sinuses*
    (6) Grooves for sigmoid sinuses*

* *Note:* These grooves are channels on the inner surface of the skull that house venous sinuses (NOT paranasal sinuses) and allow for their expansion. It is helpful to look at pictures of the dural venous sinuses in the textbook chapter on blood vessels, and observe the clear plastic skull model.

  d) Temporal
    (1) External acoustic meatus
    (2) Squamous portion
    (3) Zygomatic process
    (4) Mandibular fossa
    (5) Mastoid process (*contains air cells*)
    (6) Groove for auditory tube (Eustachianpharyngotympanic tube)- located between temporal and sphenoid as seen from an inferior view
    (7) Petrous portion
      (a) Internal acoustic meatus
    (8) Jugular foramen
    (9) Carotid canal

e) Sphenoid
    (1) Sella turcica
      (a) Hypophyseal fossa
    (2) Sphenoidal sinus
    (3) Optic foramen (canal)
    (4) Optic (chiasmatic) groove
    (5) Superior orbital fissure

f) Ethmoid
    (1) Cribriform plate
      (a) Cribriform foramina
    (2) Perpendicular plate of the ethmoid
    (3) Lateral masses
      (a) Superior nasal conchae
      (b) Middle nasal conchae
      (c) Orbital plate
      (d) Ethmoid air cells
    (4) Crista galli

g) Maxilla
    (1) Maxillary sinuses
    (2) Palatine process *(forms portion of hard palate)*
    (3) Frontal process of the maxilla
    (4) Zygomatic process of the maxilla

h) Zygomatic
    (1) Temporal process
    (2) Frontal process of the zygomatic
    (3) Maxillary process

i) Palatine
    (1) Perpendicular plate of the palatine
    (2) Horizontal plate *(forms portion of hard palate)*

j) Mandible
    (1) Condylar process
    (2) Coronoid process of the mandible

9) Know the names of the seven bones that contribute to the orbit and the specific bony structures that form the hard palate.

c. Vertebral Column

1) General features (be able to locate on all all vertebrae)
    a) Body of vertebra
    b) Vertebral arch (neural arch)
      (1) Pedicle
      (2) Lamina
    c) Spinous process
    d) Transverse process
    e) Superior articular processes
      (1) Superior articular facets
    f) Inferior articular processes
      (1) Inferior articular facets
    g) Vertebral foramen
    h) Intervertebral foramen

2) Cervical vertebrae (7)
   a) Transverse foramen
   b) Spinous process (this is often bifid)
   c) Atlas (C1)
   d) Axis (C2)
      (1) Dens
   e) Vertebral prominens - spinous process of C7

3) Thoracic vertebrae (12)
   a) Transverse costal facets
   b) Superior demifacets (visible on T2-T9)
   c) Inferior demifacets (visible on T1-T8)
   d) Costal facets (individual facets visible on T1 and T10-T12)

4) Lumbar vertebrae (5)

5) Sacrum (5 fused vertebrae)
   a) Superior articular processes
      (1) Superior articular facets
   b) Auricular surface of sacrum (forms sacroiliac joint)
   c) Body of sacrum
   d) Median sacral crest
   e) Anterior and posterior sacral foramina
   f) Sacral canal

6) Coccyx (2 to 5 fused vertebrae)

d. **The Thorax**
   The sternum, ribs, costal cartilages, and thoracic vertebrae form a cone-shaped enclosure, called the thorax.

1) Bones of the Thorax
   a) Rib
      (1) Head of rib
      (2) Tubercle of rib
      (3) Neck of rib
      (4) True ribs (first 7 pairs)
      (5) False ribs (last 5 pairs, including last 2 pairs of floating ribs)
   b) Sternum
      (1) Manubrium
         (a) Articulating facets for clavicles
         (b) Articulating facets for ribs
         (c) Suprasternal notch (jugular notch)
      (2) Body of sternum
         (a) Articulating facets for ribs
      (3) Xiphoid process

2. APPENDICULAR SKELETON
This includes the pectoral and pelvic girdles as well as the upper and lower limbs.

a. **Pectoral (shoulder) Girdle**
1) Clavicle
    a) Sternal end
    b) Acromial end
    c) Conoid tubercle
2) Scapula
    a) Spine of scapula
    b) Acromion process
    c) Superior border
    d) Medial border
    e) Lateral border
    f) Inferior angle
    g) Glenoid fossa (cavity)
    h) Supraglenoid tubercle
    i) Infraglenoid tubercle
    j) Coracoid process
    k) Supraspinous fossa
    l) Infraspinous fossa
    m) Subscapular fossa

b. **Upper Limb**
1) Humerus
    a) Head of humerus
    b) Greater tubercle
    c) Lesser tubercle
    d) Intertubercular groove (sulcus)
    e) Deltoid tuberosity
    f) Capitulum
    g) Trochlea
    h) Coronoid fossa
    i) Olecranon fossa
2) Ulna
    a) Head of ulna
    b) Olecranon process
    c) Coronoid process of ulna
    d) Trochlear notch
    e) Radial notch
    f) Styloid process of ulna
    g) Ulnar tuberosity
3) Radius
    a) Head of radius
    b) Radial tuberosity
    c) Ulnar notch
    d) Styloid process of radius

4) Hand
    a) Carpals – no need to learn individual names of these 8 bones
    b) Metacarpals – I to V
    c) Phalanges – note that the singular form is *phalanx*
        (1) proximal phalanges – I to V
        (2) middle phalanges – II to V
        (3) distal phalanges – I to V

c. **Pelvic (hip) Girdle**

The pelvic girdle is composed of 2 ossa coxae or hip bones; together with the sacrum and coccyx they form the pelvis. Each os coxae consists of an ilium, ischium and pubis. In the adult the three parts are completely fused.

1) Os Coxae
    a) Obturator foramen

    b) Acetabulum
      (1) Lunate surface
    c) Ilium
      (1) Crest
      (2) Anterior superior iliac spine
      (3) Anterior inferior iliac spine
      (4) Posterior superior iliac spine
      (5) Posterior inferior iliac spine
      (6) Greater sciatic notch
      (7) Auricular surface of ilium / os coxae (forms sacroiliac joint)

    d) Ischium
      (1) Ischial tuberosity
      (2) Ischial spine
      (3) Ischial ramus

    c) Pubis
      (1) Body of pubis
      (2) Pubic symphysis
      (3) Superior pubic ramus
      (4) Inferior pubic ramus
      (5) Pubic tubercle

d. **Lower Limb**
1) Femur
    a) Head of femur
    b) Fovea capitis
    c) Neck of femur
    d) Greater trochanter
    e) Lesser trochanter
    f) Gluteal tuberosity
    g) Linea aspera
    h) Lateral condyle of femur
    i) Medial condyle of femur
    j) Lateral epicondyle
    k) Medial epicondyle
    l) Popliteal fossa
    m) Patellar surface

2) Tibia
    a) Tibial tuberosity
    b) Lateral condyle of tibia
    c) Medial condyle of tibia
    d) Medial malleolus
    e) Soleal (popliteal) line see p. 1403 in new black Gray's Anatomy

3) Fibula
    a) Head of fibula
    b) Lateral malleolus

4) Patella

5) Foot
    a) Tarsals – there are 7; learn these 3:
        (1) Calcaneus
        (2) Talus
        (3) 1st (medial) cuneiform
    b) Metatarsals – I to V
    c) Phalanges
        (1) proximal phalanges – I to V
        (2) middle phalanges – II to V
        (3) distal phalanges – I to V

FYI – Interesting information about the Pelvis

Bones and Biological Sex: The formation and development of bone is influenced by sex hormones and many bones, especially the pelvis, reflect differences associated with childbearing. These differences in biologically male and female skeletal anatomy are extremely useful in fields such as anthropology, forensic medicine, and of course human anatomy. The chart shown below provides a comparison of the male and female pelvis. We will not be determining the sex of the pelvises in lab, because even using more than one character, you may be fooled!

Comparison of Male and Female Bony Pelvises

Features	Male	Female
Overall	Narrow, heavy, compact	Wide, light, capacious
Distance between anterior superior iliac spines	Narrow — thus narrower hips	Wider — thus wider hips
Inlet of true (lesser) pelvis	Heart–shaped; cavity deep and narrow	Almost perfect circle; cavity shallow and wide
Angle of pubic arch	Acute	Obtuse
Sacrum	Long, narrow, curved	Short, broad, straighter
Ischial spines	Sharp, directed inward	Blunt, not directed as far inward
Outlet of true (lesser) pelvis	Small; coccyx rather fixed	Large; coccyx rather flexible
Sciatic notches	Narrow, deep	Wide, shallow
Obturator foramina	Large, rather rounded	Small, rather triangular
Acetabula	Large, directed forward	Small, directed outward

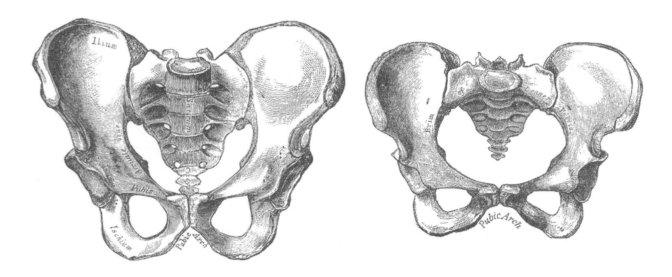

CHAPTER 6

Muscular System

Muscular System

I. Muscle Tissue: Models

A. On the model of a skeletal muscle fiber identify and locate the following:
1. Endomysium
2. Sarcolemma
3. Neuromuscular junction
   a. Axon terminal (synaptic knob) of motor neuron
   b. Synaptic cleft
   c. Motor end plate
4. Sarcoplasm
5. Nuclei
6. Myofibrils
7. Sarcomeres

B. On the model of myofibrils note the following:
1. Myofibrils
2. Thick and Thin Myofilaments
3. Sarcoplasmic Reticulum
4. T-tubules (transverse tubules)
5. Sarcomeres
6. A Bands, H Bands, I Bands, Z Lines, M Lines
7. Glycogen granules
8. Mitochondria
9. Sarcolemma

II. Histology:

A. **Skeletal Muscle** – use both longitudinal and cross section of this tissue and locate the following:
1. Epimysium
2. Perimysium
3. Endomysium
4. Fascicles
5. Fibers (note cylindrical shape)
6. Nuclei
7. Striations
8. Myofibrils (can also be seen in x.s. as dark points within the muscle fiber)

B. **Neuromuscular Junction**
Examine this section and locate the following:
1. Axon of motor neuron
2. Neuromuscular junction

C. **Smooth Muscle**

Small intestine will be used to examine smooth muscle; this organ has 2 layers of smooth muscle in the outer layers of the organ.

Note the following **characteristic features** of smooth muscle tissue:

- spindle-shaped fibers
- single central nucleus
- absence of striations

D. **Cardiac Muscle**

Examine this tissue in both cross section & longitudinal views.

Note the following **characteristic features** of cardiac muscle tissue:
- branching fibers
- central nuclei
- striations
- intercalated discs

III. Muscles as Organs: Gross Anatomy

Using models, charts and cadavers you must learn the 63 muscles listed on page 39. These muscles will be isolated during dissections. Information about dissections is in Chapter 14.

Learn the following muscle groups:
Rotator cuff: supraspinatus, infraspinatus, subscapularis, teres minor
Quadriceps: rectus femoris, vastus lateralis, vastus medialis, vastus intermedius
Hamstrings: biceps femoris, semimembranosus, semitendinosus

IV. Knee Joint

Remember that even though only muscles have actions, you must still know the functions of these structures:

A. Lateral meniscus
B. Medial meniscus
C. Posterior cruciate ligament
D. Anterior cruciate ligament
E. Fibular (lateral) collateral ligament
F. Tibial (medial) collateral ligament
G. Transverse ligament

Note: As you learn the actions of muscles note that in anatomy, what is called the *arm* in common parlance is instead the *upper limb*. The upper limb consists of the *arm* (proximally) and *forearm* (distally). Similarly, what is commonly called the *leg is* is the *lower limb* in anatomy. The lower limb consists of the *thigh* (proximally) and *leg* (distally).

Region 1: **Head**

    Temporalis
    Masseter
    Frontalis
    Buccinator
    Orbicularis Oculi
    Orbicularis Oris

Region 2: **Neck, Back**

    Sternocleidomastoid
    Trapezius
    Levator Scapulae
    Rhomboid Major
    Rhomboid Minor
    Latissimus Dorsi

Region 3: **Posterior Upper Limb**

    Deltoid
    Supraspinatus
    Infraspinatus
    Subscapularis
    Teres Minor
    Teres Major
    Triceps Brachii (three heads)

Region 4: **Anterior Upper Limb**

    Pectoralis Major
    Pectoralis Minor
    Serratus Anterior
    Coracobrachialis
    Biceps Brachii (two heads)
    Brachialis

Region 5: **Forearm**

    Brachioradialis
    Extensor Carpi Radialis Longus
    Extensor Carpi Radialis Brevis
    Extensor Digitorum
    Extensor Carpi Ulnaris
    Flexor Carpi Ulnaris
    Palmaris Longus
    Flexor Carpi Radialis

Region 6: **Thorax, Abdomen**

    Diaphragm
    External Intercostals
    Internal Intercostals
    Rectus Abdominis
    External Obliques
    Internal Obliques
    Transverse Abdominis
    Inguinal Ligament
    Psoas Major

Region 7: **Anterior and Medial Thigh**

    Rectus Femoris
    Vastus Lateralis
    Vastus Medialis
    Vastus Intermedius
    Sartorius
    Gracilis
    Pectineus
    Adductor Longus
    Adductor Brevis
    Adductor Magnus

Region 8: **Gluteal Region and Posterior Thigh**

    Gluteus Maximus
    Gluteus Medius
    Biceps Femoris (two heads)
    Semitendinosus
    Semimembranosus

Region 9: **Leg**

    Plantaris
    Popliteus
    Gastrocnemius
    Soleus
    Tibialis Anterior
    Peroneus (Fibularis) Longus

Region 10: **Knee Joint**

    Lateral Meniscus
    Medial Meniscus
    Anterior Cruciate Ligament
    Posterior Cruciate Ligament
    Fibular (Lateral) Collateral Ligament
    Tibial (Medial) Collateral Ligament
    Transverse Ligament

Bold–face type denotes muscles for which origin, insertion & action(s) are learned; for others learn name & action(s).

MUSCLE	ORIGIN	INSERTION	ACTION
Deltoid	Clavicle, acromion process, and spine of scapula	Deltoid tuberosity	Abduct arm
Supraspinatus	Supraspinous fossa	Superior part of greater tubercle	Abduct arm
Infraspinatus	Infraspinous fossa	Middle part of greater tubercle	Laterally rotate arm
Subscapularis	Subscapular fossa	Lesser tubercle	Medially rotate arm
Teres minor	Lateral border of scapula	Inferior part of greater tubercle	Laterally rotate arm
Teres major	Inferior angle of scapula	Distal to lesser tubercle	Extend adduct, medially rotate arm
Triceps brachii	**Long head** – infraglenoid tubercle **Lateral head**– lateral and posterior diaphysis of humerus **Medial head**– posterior diaphysis of humerus	Olecranon process	Extend forearm
Coracobrachialis	Coracoid process	Medial surface of humerus	Flex, adduct arm
Biceps brachii	**Long head**–supraglenoid tubercle **Short head**– coracoid process	Radial tuberosity	Flex forearm; supinate radius
Brachialis	Anterior diaphysis of humerus	Ulnar tuberosity and coronoid process of ulna	Flex forearm
Rectus femoris	Anterior inferior iliac spine	Tibial tuberosity	Flex thigh; extend leg
Vastus lateralis	Greater trochanter and linea aspera	Tibial tuberosity	Extend leg
Vastus medialis	Linea aspera	Tibial tuberosity	Extend leg
Vastus intermedius	Anterior and lateral diaphysis of femur	Tibial tuberosity	Extend leg
Adductor longus	Anterior body of pubis	Linea aspera	Adduct thigh
Adductor brevis	Inferior pubic ramus	Linea aspera	Adduct thigh
Adductor magnus	Inferior pubic ramus, ischial ramus, and ischial tuberosity	Linea aspera	Adduct, flex thigh
Biceps femoris	**Long head**–ischial tuberosity **Short head**–linea aspera	Lateral condyle of tibia and head of fibula	Extend thigh; flex leg
Semitendinosus	Ischial tuberosity	Proximal medial diaphysis of tibia	Extend thigh; flex leg
Semimembranosus	Ischial tuberosity	Medial condyle of tibia	Extend thigh; flex leg
Gastrocnemius	Posterior side of medial and lateral condyles of femur	Calcaneus	Flex leg; plantar flex foot
Soleus	Posterior fibula, and soleal line	Calcaneus	Plantar flex foot
Tibialis anterior	Lateral condyle and diaphysis of tibia	First metatarsal and first cuneiform	Dorsiflex, invert foot

MUSCLE	ACTION
Temporalis	elevate, retract mandible
Masseter	elevate, protract mandible
Frontalis	elevate eyebrows
Buccinator	compress cheeks, assist in mastication (keeps food between teeth)
Orbicularis oculi	close eyes (squint, blink)
Orbicularis oris	close, protrude lips

MUSCLE	ACTION
Sternocleidomastoid	flex neck; rotate head to opposite side
Trapezius	elevate, adduct, stabilize scapula; extend neck
Levator scapulae	elevate scapula
Rhomboid major	retract, adduct scapula
Rhomboid minor	retract, adduct scapula
Latissimus dorsi	extend, adduct, medially rotate arm

MUSCLE	ACTION
Pectoralis major	flex, adduct, medially rotate arm
Pectoralis minor	depress scapula
Serratus anterior	abduct, protract scapula

MUSCLE	ACTION
Brachioradialis	flex forearm; supinate, pronate radius (return to relaxed position)
Extensor carpi radialis longus	extend, abduct wrist
Extensor carpi radialis brevis	extend, abduct wrist
Extensor digitorum	extend fingers
Extensor carpi ulnaris	extend, adduct wrist
Flexor carpi ulnaris	flex, adduct wrist
Palmaris longus	flex wrist
Flexor carpi radialis	flex, abduct wrist

MUSCLE	ACTION
Diaphragm	inspiration
External intercostals	elevate ribs during inhalation
Internal intercostals	depress ribs during forced exhalation
Rectus abdominis	flex vertebral column; compress abdomen
External obliques	flex, laterally flex, rotate vertebral column
Internal obliques	laterally flex, rotate vertebral column; compress abdomen
Transverse abdominis	compress abdomen
Psoas major	flex vertebral column; flex thigh

MUSCLE	ACTION
Sartorius	flex, laterally rotate thigh; flex leg
Gracilis	flex, adduct thigh; flex leg
Pectineus	flex, adduct thigh
Gluteus maximus	extend, laterally rotate thigh
Gluteus medius	abduct thigh
Plantaris	plantar flex foot
Popliteus	medially rotate tibia (unlock knee)
Peroneus (fibularis) longus	plantar flex, evert foot

FYI – A CHART TO COMPARE THREE TYPES OF MUSCLE FIBERS*

	SMOOTH	SKELETAL	CARDIAC
pseudonyms	involuntary, non–striated, unstriped, plain	voluntary, striated, striped, red and white skeletal	heart
myofibrils	inconspicuous	conspicuous	fairly conspicuous
length of fiber	0.02 mm to 0.5 mm	1 to 40 mm	0.08 mm or less
diameter of fiber	to 10 μ at thickest part	10 to 40 μ	15 μ approx
branching of fiber	none	none	frequent
composition of fiber	single cell	multinucleate syncytium	single branching cell
nucleus	central of each fiber	many nuclei at periphery	central, large
cross striations	absent	present	present
intercalated discs	absent	absent	present
contraction	slow, rhythmic, sustained	rapid, powerful; not sustained	moderately rapid, with rests between contractions; not sustained
control of contraction	impulses from CNS not essential for contraction	neurogenic; contracts only in response to motor impulses from CNS	myogenic; but rate controlled by autonomic nervous system
distribution	alimentary, respiratory, and urogenital tracts; blood vessel and larger lymphatics; main ducts of glands; ciliary muscle of eye; arrector pili muscle of skin	locomotory muscles; sheets of muscle of abdominal wall, diaphragm	heart only

*This chart is provided for your organizational reference only

CHAPTER 7

Coelom and Viscera, Cardiovascular and Lymphatic Systems

COELOM, VISCERA, AND CIRCULATORY SYSTEMS

This part of the lab covers the cardiovascular and lymphatic systems as well as a more in-depth study of the anterior body cavity or coelom, and the visceral organs and serous membranes that fill this cavity. Dissections will continue in this part of the lab. The second set of student dissectors will reveal the listed visceral organs and blood vessels.

This list below outlines the major structures (bold faced) to be learned in this part of the lab.

coelom	the ventral body cavity; including all of its major subdivisions
diaphragm	the anatomical structure which divides the coelom into, the thoracic and abdominopelvic cavities
viscera	the names of all organs found in the coelom (details about these organs will be studied in subsequent chapters)
membranes	all serous membranes associated with the coelom
blood vessels	histology of arteries, veins and capillaries; names of major blood vessels; venous valves (shown on femoral vein)
heart	gross and histological features
blood	identification of all formed elements
lymphatic system	histological examination of lymph node; identification of lymphatic vessels on charts only
nerves	three nerves are dissected during this part of the lab (phrenic, vagus, and recurrent laryngeal); you must be able to identify these in the cadaver on the nervous system laboratory exam
spermatic cord	dissect now, identify on reproductive system laboratory exam

Before dissections begin, it is important for students to observe the following undisturbed structures:

1. **Falciform ligament** attaches liver to the anterior body wall
   a) **Ligamentum teres** - remnant of umbilical vein
2. **Coronary ligament** attaches liver to diaphragm.
3. **Greater omentum** attaches stomach to transverse colon
4. **Lesser omentum** attaches stomach to liver
5. **Mesentery proper** attaches jejunum and ileum to posterior body wall
6. **Mesocolon** attaches transverse colon and sigmoid colon to posterior body wall

I. Coelom and Viscera

A. Coelom
Identify the boundaries of the coelom (ventral body cavity), its major subdivisions, and their contents.

1. Cavities
   a. **thoracic cavity** includes **pleural cavities, pericardial cavity, mediastinum**
   b. **abdominopelvic cavity** includes **abdominal cavity** and **pelvic cavity**
   c. **peritoneal cavity** is a subset of the abdominopelvic cavity. The peritoneal cavity is filled with serous fluid and surrounds all the abdominopelvic viscera that are **intraperitoneal**

2. Serous membranes – know the specific tissues that form all serous membranes. Identify each of the serous membranes in the body cavities, distinguishing between **visceral** and **parietal** layers where appropriate (a, b, c).
   a. peritoneum   d. mesentery
   b. pleura   e. ligament
   c. pericardium   f. omentum

3. Diaphragm – Identify where the following structures pass through the diaphragm.
   a. aorta - aortic hiatus   c. esophagus - esophageal hiatus
   b. inferior vena cava - caval haitus d. vagus nerves - esophageal hiatus

4. Note the structures that enter the coelom via the **thoracic inlet** (esophagus, trachea, major arteries and veins).

5. Note the structures that exit the coelom via the **femoral canal** (femoral A, V, L, N) and via the **inguinal canal** (spermatic cord in males, round ligament in females)

B. Viscera
1. **Retroperitoneal organs** - Identify those organs that are retroperitoneal; i.e., those located "behind" the parietal peritoneum:

duodenum	adrenal glands	aorta
pancreas	kidneys	inferior vena cava
ascending colon	ureters	
descending colon	urinary bladder	

2. **Intraperitoneal organs** – Identify those organs that are intraperitoneal; i.e., those wrapped in visceral peritoneum and surrounded by the peritoneal cavity:

jejunum	stomach
ileum	liver
transverse colon	gallbladder
sigmoid colon	spleen

II Cardiovascular System

The human body has two systems that circulate fluid: the cardiovascular system and the lymphatic system. The parts of the cardiovascular system are the blood vessels, the blood, and the heart.

A. Heart (gross anatomy)

Examine the following on human hearts as well as models and charts:

1. Pericardial sac
   a. fibrous pericardium
   b. parietal pericardium – note that this is a serous membrane
2. Heart wall
   a. epicardium (visceral pericardium) – note that this is a serous membrane
   b. myocardium
   c. endocardium
3. External features
   a. apex
   b. coronary sulcus
   c. interventricular sulcus (anterior and posterior)
4. Right atrium and left atrium (atria, pl.)
   a. right and left auricles
   b. pectinate muscles (in right atrium and both auricles)
   c. interatrial septum
5. Atrioventricular valves
   a. right artioventricular valve (tricuspid value)
   b. left atrioventricular valve (mitral, bicuspid valve)
6. Ventricles, right and left
   a. chordae tendineae
   b. trabeculae carneae
   c. papillary muscles
   d. interventricular septum
7. Semilunar valves
   a. pulmonary semilunar valve
   b. aortic semilunar valve
8. Coronary vessels
   a. coronary arteries, right and left
   b. anterior and posterior interventricular arteries
   c. cardiac veins
   d. coronary sinus
9. Remnants of embryonic structures
   a. fossa ovalis (remnant of foramen ovale)
   b. ligamentum arteriosum (remnant of ductus arteriosus)
10. Great vessels
    a. superior vena cava
    b. inferior vena cava
    c. ascending aorta
    d. pulmonary trunk (leads to right and left pulmonary arteries)
    e. superior pulmonary veins, right and left
    f. inferior pulmonary veins, right and left

B. Blood Vessels (gross anatomy)

Learn names and location of these selected vessels using cadavers, models, and charts. *
Named vessels listed as plural are paired (right and left)

Arteries

Pulmonary Trunk

    Pulmunary Arteries – *deox blood to lungs*

Aorta: ascending, arch, thoracic, abdominal

Brachiocephalic Artery – *head, neck, R. upper limb*

Common Carotid Arteries – *head, neck*

    Internal Carotid Arteries – *brain*

    External Carotid Arteries – *face*

Vertebral Arteries – *brain* **

Subclavian Arteries – *head, neck, upper limbs*

Thyrocervical Arteries (Trunks) – *thyroid, neck*

Internal Thoracic Arteries – *ant. thoracic wall*

Celiac Artery (Trunk)

    Gastric Artery – *stomach*

    Splenic Artery – *spleen*

    Common Hepatic Artery – *liver*

Renal Arteries – *kidneys*

    Suprarenal Arteries – *adrenal glands*

Superior Mesenteric Artery – *sm & proximal lg intestine, pancreas*

Ovarian / Testicular Arteries – *gonads*

Inferior Mesenteric Artery – *distal large intestine*

Common Iliac Arteries

    Internal Iliac Arteries – *pelvis*

    External Iliac Arteries – *lower limbs*

    Femoral Arteries – *lower limbs*

Veins

Pulmonary Veins, superior and inferior – *drain ox blood from lungs*

Superior & Inferior Vena Cava

Brachiocephalic Veins – *head, neck, upper limbs*

Internal Jugular Veins – *brain*

External Jugular Veins – *face*

Vertebral Veins – *brain*

Subclavian Veins – *head, neck, upper limbs*

Cephalic Veins – *upper limbs*

Azygos Vein – *thoracic body wall*

Hepatic Veins – *liver*

Hepatic Portal Vein

    Gastric Vein – *stomach*

    Splenic Vein – *spleen*

    Superior Mesenteric Vein – *small and proximal large intestine, pancreas*

    Inferior Mesenteric Vein – *distal large intestines*

Renal Veins – *kidneys*

    Suprarenal Veins – *adrenal glands*

Ovarian / Testicular Veins – *gonads*

Common Iliac Veins

    Internal Iliac Veins – *pelvis*

    External Iliac Veins – *lower limbs*

Femoral Veins – *lower limbs*

    venous value

Saphenous Veins – *medial thighs*

Note 1: On lab practical exam you must indicate whether a vessel is an artery (A) or vein (V) and for each vessel know the region supplied or drained – given in italics on above list.

**Note 2:* The vertebral, thyrocervical and internal thoracic arteries originate from the subclavian artery very close together. There can be variation in the order; usually the vertebral artery is more medial than the thyrocervical artery.

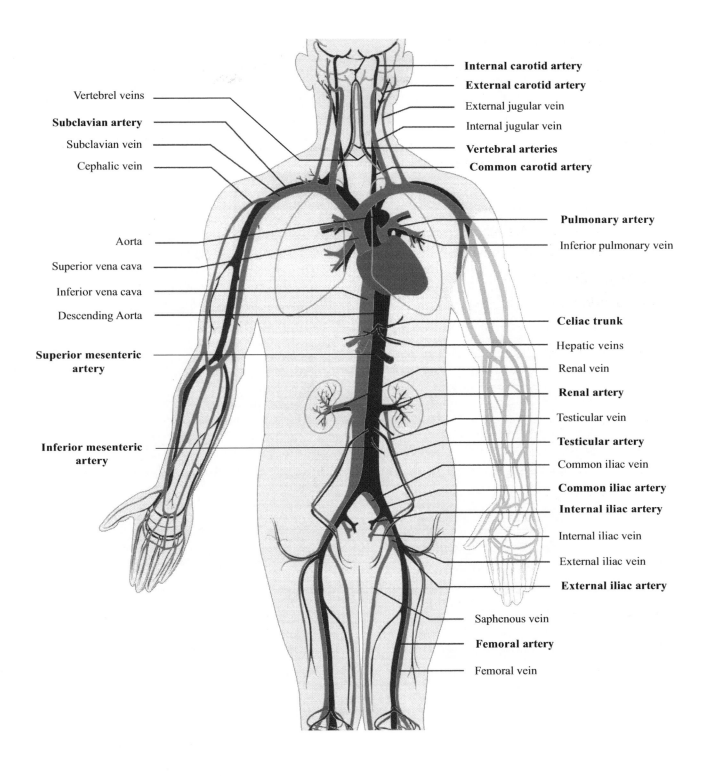

Vertebrel veins

Subclavian artery

Subclavian vein

Cephalic vein

Aorta

Superior vena cava

Inferior vena cava

Descending Aorta

Superior mesenteric artery

Inferior mesenteric artery

Internal carotid artery

External carotid artery

External jugular vein

Internal jugular vein

Vertebral arteries

Common carotid artery

Pulmonary artery

Inferior pulmonary vein

Celiac trunk

Hepatic veins

Renal vein

Renal artery

Testicular vein

Testicular artery

Common iliac vein

Common iliac artery

Internal iliac artery

Internal iliac vein

External iliac vein

External iliac artery

Saphenous vein

Femoral artery

Femoral vein

C. **Blood Vessels (histology)**

Examine slides of blood vessels. Identify the following structures found in the walls of blood vessels. Distinguish between an **artery** and a **vein**, and between a **muscular** and an **elastic artery.** Compare your slides with the model of the vein and artery. See pages 51 and 52 for helpful hints.

    1. Elastic Artery: aorta
        a. tunica interna: endothelium
        b. tunica media: elastic CT & smooth muscle tissue
        c. tunica externa: white fibrous / areolar CT

    2. Muscular Artery:
        a. tunica interna: endothelium & internal elastic lamina
        b. tunica media: smooth muscle tissue(few elastic fibers can be visible)
        c. tunica externa: external elastic lamina and white fibrous / areolar CT

    3. Vein
        a. tunica interna: endothelium
        b. tunica media: smooth muscle tissue
        c. tunica externa: white fibrous / areolar CT

    4. Vasa Vasorum: (small vessels that supply blood to large vessels,
                           (best seen on aorta slide)

D. **Blood**

Examine blood smear slides. Identify the structures and functions of the formed elements of the blood:

    1. Erythrocytes (red blood cells)

    2. Leukocytes (white blood cells)
        a. granulocytes
            1) Neutrophil
            2) Eosinophil
            3) Basophil

        b. agranulocytes
            1) Lymphocyte
            2) Monocyte

    3. Platelets

E. **Red Bone Marrow**

Examine red bone marrow slides. There are many different cells present, but only one you need to identify:
    1. Megakaryocyte

III. Lymphatic System

A. Gross anatomical structures

1. Thoracic duct - note pattern of drainage (*use charts*)
2. Right lymphatic duct - note pattern of drainage (*use charts*)
3. Cisterna chyli (*use charts*)
4. Lymph nodes
5. Spleen
6. Thymus (*no longer present in adults*, *use charts*)
7. Appendix

B. Histology

Lymph Node Identify and explain the function of the following parts:

1. Capsule
2. Cortex
   a. Cortical sinuses
   b. Trabeculae
   c. Lymphatic follicles (nodules), each with a germinal center
3. Medulla
   a. Medullary sinuses
   b. Medullary cords (lymphatic cords)

Some General Differences Between Distributing Arteries and Their Companion Veins

Arteries	Veins
Have a smaller overall diameter; smaller lumen	Have a larger overall diameter; larger lumen
Thicker walls, to withstand pressure	Thinner walls, less pressure
The thicker walls do not collapse after death	The thinner walls collapse if blood is drained out; may appear flattened in sections
Tunica media is a much thicker muscular layer	Tunica media is a thin, muscular layer
Internal & external elastic laminae are both well developed	Elastic laminae less well developed
Tunica externa is half the thickness of the media, has high elastin content	Tunica externa is thickest layer of wall, composed of collagen, few elastic fibers
Have no valves	Have valves in extremities to prevent backflow

Histological Comparison of Cross-Sections of Arteries, Veins, and Nerves

A. Arteries

The walls are thick, and the lumen is large and circular. The lumen is lined with flat squamous cells, each of which present a dark nucleus that bulges into the lumen. The large lumen may (or may not) contain blood cells which are clustered in irregular masses and look like bits of orange–red or clear broken glass.

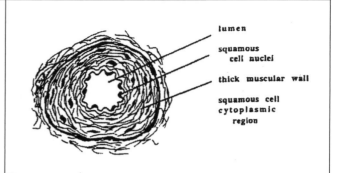

B. Veins

Veins are similar to arteries but with a larger, highly irregular, and smooth lumen and thinner walls. The lumen is lined with flat squamous cells and like arteries it may or may not contain blood cells. If it were full of blood when the section was made the lumen could be solid but appear opaque. If the lumen is empty it will appear clear.

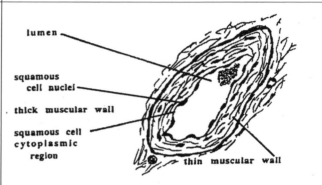

C. Nerves

Nerves are constructed like electrical cables (which they are) where the cable contains many (100s–1000s) insulated wires (axons) inside. Nerves are circular and they lack a lumen; they are filled with axons. In cross sections the walls are thinner than ducts and the central area is packed uniformly with dark dots (axons) each surrounded by white coverings of insulating myelin. Occasionally the internal bundles will be separated by bands of connective tissue.

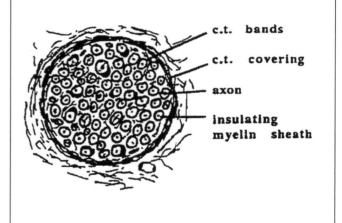

CHAPTER 8

Nervous System

NERVOUS SYSTEM

I. Histology:

A. Identify the following on the **multipolar neuron** model:
1. Cell body
    a. Nucleus
    b. Nucleolus
    c. Nissl bodies
2. Processes
    a. Axon
        1) Axon hillock
        2) Synaptic knob
    b. Dendrites
3. Schwann cell
    a. Neurilemma
    b. Myelin sheath
4. Nodes of Ranvier

B. Identify the following on the **neuroglia** slide:
1. Astrocytes
    a. Cell body
    b. Processes
        1) Perivascular feet

C. Identify the following on the **nerve slides** (*study both longitudinal and cross sections*):
1. Neuron fibers (axons)
2. Schwann cells
    a. Myelin sheath (*stains black with osmium on some slides*)
    b. Neurilemma
3. Nodes of Ranvier
4. Endoneurium

D. Identify the following on the **dorsal root ganglion** slide:
1. Unipolar neurons
    a. Cell body
        1) Nucleus
        2) Nucleolus
        3) Nissl bodies
    b. Axons
2. Satellite cells

E. Identify the following on both models and slides of the **spinal cord:**
1. Posterior median sulcus
2. Anterior median fissure
3. Central canal
    a. Ependymal cells
4. Gray matter
    a. Posterior (dorsal) horn
    b. Anterior (ventral) horn
    c. Lateral horn (*where applicable*)
    d. Gray commissure
    e. Multipolar neurons
        1) Cell body
            a. Nucleus
            b. Nucleolus
            c. Nissl bodies
        3) Processes (*no need to distinguish axons from dendrites*)
5. White matter
    a. Posterior (dorsal) funiculi (columns)
    b. Anterior (ventral) funiculi (columns)
    c. Lateral funiculi (columns)

II. Gross Anatomy:

Use charts, models, specimens, and cadavers to identify the following structures:

A. **Spinal cord** and associated structures (*all structures are visible on models; those marked* * *are also visible on prosections*)

1. Cervical and lumbosacral enlargements*
2. 31 pairs of spinal nerves and their divisions (cervical, thoracic, etc.)
3. Gray matter areas in x.s. (dorsal, ventral, lateral horns)
4. White matter areas in x.s. (dorsal, ventral, lateral funiculi)
5. Meningeal spaces and coverings
    - a. epidural space
    - b. dura mater *
    - c. subdural space
    - d. arachnoid
    - e. subarachnoid space
    - f. pia mater *
6. Denticulate ligaments *
7. Conus medullaris *
8. Cauda equina *
9. Terminal filament of pia * (filum terminale interna)
10. Posterior median sulcus
11. Anterior median fissure *
12. Anterior (ventral) and posterior (dorsal) roots and rootlets *
13. Posterior (dorsal) root ganglion
14. Anterior (ventral) ramus, posterior (dorsal) ramus
15. Sympathetic trunk ganglia

B. **Brain**

1. Cerebrum (*derived from the telencephalon*)
    - a. Cerebral hemispheres
        1) Cerebral cortex (*comprised of gray matter*)
        2) Longitudinal fissure
        3) Frontal lobe
        4) Parietal lobe
        5) Occipital lobe
        6) Temporal lobe
        7) Insula
        8) Gyri: precentral gyrus and postcentral gyrus
        9) Sulci: central sulcus and lateral sulcus
        10) White matter (*note discussion of neural pathways in text book*)
            - a) Corpus callosum
            - c) Corona radiata
            - b) Fornix
            - d) Internal capsule
        11) Lateral ventricles (left and right)
            - a) Choroid plexus
        12) Septum pellucidum

b. Cerebral/Basal nuclei
　1) Caudate
　2) Lentiform
　　a) Putamen
　　b) Globus pallidus
c. Hippocampus
d. Amygdala
e. Cranial nerve I (olfactory nerve) and olfactory bulb

2. Diencephalon *(derived from the diencephalon)*
　a. Thalamus
　　1) Interthalamic adhesion (intermediate mass)
　b. Hypothalamus
　　1) Infundibulum
　　2) Pituitary
　　3) Mammillary body
　c. Epithalamus – pineal gland
　d. Interventricular foramen
　e. Third ventricle
　　1) Choroid plexus
　f. Cranial nerve II (optic nerve) and optic chiasm

3. Midbrain *(derived from the mesencephalon)*
　a. Cerebral peduncles
　b. Cerebral aqueduct (aqueduct of Sylvius)
　c. Corpora quadrigemina
　　1) superior colliculi
　　2) inferior colliculi
　d. Substantia nigra
　e. Red nucleus
　f. Cranial nerve III (oculomotor nerve)
　g. Cranial nerve IV (trochlear nerve)

4. Pons *(derived from the metencephalon)*
　a. Cerebellar peduncles (superior, middle, inferior)
　b. Cranial nerve V (trigeminal nerve)
　c. Cranial nerve VI (abducens nerve)
　d. Cranial nerve VII (facial nerve)
　e. Cranial nerve VIII (vestibulocochlear nerve)

5. Cerebellum *(derived from the metencephalon)*
　a. Lateral cerebellar hemispheres
　b. Vermis
　c. Folia
　d. Arbor vitae *(comprised of white matter)*

6. Medulla oblongata *(derived from the myelencephalon)*
    a. Olives
    b. Pyramids, decussation of pyramids
    c. Fourth ventricle
        1) Choroid plexus
        2) Median aperture
        3) Lateral apertures
    d. Cranial nerve IX (glossopharyngeal nerve)
    e. Cranial nerve X (vagus nerve) *(also view on cadaver)*
        1) Recurrent laryngeal nerve *(view on cadaver)*
    f. Cranial nerve XI (accessory nerve)
    g. Cranial nerve XII (hypoglossal nerve)

7. Meninges and spaces *(listed in order)*
    a. Epidural space *(around spinal cord, but not brain)*
    b. Dura mater
        1) Falx cerebri
            a) Superior sagittal sinus
                i. Arachnoid villi
            b) Inferior sagittal sinus
        2) Falx cerebelli
            a) Occipital sinus
        3) Tentorium cerebelli
            a) Straight sinus
            b) Transverse sinuses
            c) Sigmoid sinuses
        4) Confluence of sinuses
    c. Subdural space *(a potential space; view on model)*
    d. Arachnoid
    e. Subarachnoid space
    f. Pia mater

8. Circle of Willis and major associated arteries
    a. Internal carotid arteries
    b. Vertebral arteries
    c. Basilar artery
    d. Anterior cerebral arteries
    e. Middle cerebral arteries
    f. Posterior cerebral arteries
    g. Anterior communicating artery
    h. Posterior communicating arteries

C. **Peripheral Nervous System**
   1. Cranial nerves – 12 pairs (*refer to chart, pages 65–66; note vagus and recurrent laryngeal nerves in cadaver*)
   2. Spinal nerves – 31 pairs (*visible on models*)
      a. Posterior ramus
      b. Anterior ramus
   3. Cervical plexus
      a. Phrenic nerve
   4. Brachial plexus
      a. Radial nerve
      b. Median nerve
      c. Ulnar nerve
   5. Lumbar plexus
      a. Femoral nerve
   6. Sacral plexus
      a. Sciatic nerve
         1) Tibial nerve
         2) Common fibular (peroneal) nerve
   7. ANS – autonomic nervous system
      a. Gray communicating ramus
      b. White communicating ramus
      c. Autonomic ganglia
         1) Sympathetic trunk/chain (paravertebral) ganglia
         2) Collateral (prevertebral) ganglia (*occasionally seen in cadavers*)
         3) Terminal ganglia (*can only be seen histologically and will be studied with the digestive system*)

GENERAL SUBDIVISIONS OF THE BRAIN

GENERAL DIVISION	MAJOR STRUCTURES INCLUDED

Prosencephalon
(forebrain)

— Telencephalon
      cerebral hemispheres
      olfactory bulbs
      basal nuclei
      lateral ventricles
      projection fibers: internal capsule, corona radiata
      commissural fibers: corpus callosum

— Diencephalon
      epithlamus
      thalamus
      hypothalamus
      pituitary gland
      pineal body
      third ventricle

Mesencephalon (midbrain)
      corpora quadrigemina
      cerebral peduncles
      cerebral aqueduct
      substantia nigra
      red nucleus

Rhombencephalon
(hindbrain)

— Metencephalon
      pons
      cerebellum
      part of fourth ventricle

— Myelencephalon
      medulla oblongata
      part of fourth ventricle

CRANIAL NERVES

NO.	NAME	FUNCTIONS	ORIGIN*	EXIT FORAMINA*
I	olfactory	**Sensory**: olfaction	cells of nasal mucosa (olfactory bulb, tract to temporal lobe)	cribriform foramina
II	optic	**Sensory**: vision	ganglion cells of retina (optic tract to thalamus, occipital lobe)	optic foramen
III	oculomotor	**SNS Motor⁺**: eye muscle control **Parasympathetic:** ciliary muscles (focusing), sphincter layer of iris (constrict pupil)	midbrain	superior orbital fissure
IV	trochlear	**SNS Motor**: eye muscle control	midbrain (posterior)	superior orbital fissure
V	trigeminal		pons (lateral)	
	*ophthalmic branch	**Sensory**: touch, pain, temperature, muscle sense	sensory receptors, face	superior orbital fissure
	*maxillary branch	**Sensory**: touch, pain	sensory receptors	foramen rotundum
	*mandibular branch	**SNS Motor**: muscles of mastication **Sensory**: touch, pain temperature, muscle sense	pons, sensory receptors, lower jaw	foramen ovale
VI	abducens	**SNS Motor**: eye muscle control	pons	superior orbital fissure
VII	facial	**SNS Motor**: facial expression **Parasympathetic**: lacrimal & salivary gland secretion **Sensory**: taste	pons, taste buds	stylomastoid foramen
VIII	vestibulocochlear:			
	*vestibular	**Sensory**: equilibrium	semicircular canals, saccule, utricle (to pons, cerebellum)	internal auditory meatus
	*cochlear	**Sensory**: hearing	cochlea (to medulla, thalamus, temporal lobe)	internal auditory meatus

*FYI only; not for lab exam

Note⁺: all nerves carrying motor innervation also carry proprioceptive or muscle sense sensory information

SNS motor = somatic motor

Parasympathetic = autonomic motor, parasympathetic (there is no sympathetic output in cranial nerves)

CRANIAL NERVES

NO.	NAME	FUNCTIONS	ORIGIN*	EXIT FORAMINA*
IX	glossopharyngeal	**SNS Motor**: pharynx, swallowing **Parasympathetic**: salivary gland secretion **Sensory**: general sense, taste, blood pressure	medulla oblongata, taste buds	jugular foramen
X	vagus+	**SNS Motor**: pharynx & larynx (swallowing & speech) **Parasympathetic**: control of heart, smooth muscle of airways, smooth muscle & glands of GI tract **Sensory**: visceral sense	medulla oblongata	jugular foramen
XI	accessory	**SNS Motor**: swallowing, movement of head & neck	medulla oblongata & cervical spinal cord	jugular foramen
XII	hypoglossal	**SNS Motor**: tongue (swallowing & speech)	medulla oblongata	hypoglossal canal

* FYI only; not for lab exam

+ Recurrent laryngeal nerve is a branch of vagus nerve; must be isolated and learned; innervates larynx (intrinsic muscles and sensation from mucosa)

NAMED NERVES (derived from spinal nerves)

Name   **Innervates (sensory and motor functions)**

phrenic — diaphragm

radial — posterior muscles of arm, forearm (e.g. triceps brachii, brachioradialis, extensors)

median — wrist flexors of lateral forearm (e.g. palmaris longus, flexor carpi radialis)

ulnar — wrist flexors of medial forearm (e.g. flexor carpi ulnaris)

femoral — muscles of anterior thigh (e.g. quads, sartorius)

sciatic

    tibial — posterior muscles of thigh, leg (e.g. hamstrings, gastrocnemius, soleus)

    common fibular (peroneal) — anterior and lateral muscles of leg (e.g. tibialis anterior, fibularis (peroneus) longus)

CHAPTER 9

Special Senses

SPECIAL SENSES

I. Gross Anatomy:

A. Eye:

Locate and identify the following *(all structures are visible on models of the eye; those marked * are also visible on prosections)*:

1. Fibrous tunic
   a. Sclera *
   b. Cornea *
2. Vascular tunic
   a. Choroid
   b. Ciliary body *(no need to distinguish ciliary process and ciliary muscle)*
   c. Iris *
   d. Pupil *
3. Nervous tunic - retina
   a. Macula lutea
      1) Fovea centralis
   b. Optic disc
   c. Optic nerve *
4. Internal Chambers and Fluids
   a. Anterior cavity, filled with aqueous humor
      1) Posterior chamber
      2) Anterior chamber
      3) Scleral venous sinus (canal of Schlemm)
   b. Posterior (vitreous) cavity, filled with vitreous humor
5. Lens *
6. Suspensory ligaments
7. Conjunctiva
8. Extrinsic eye muscles
   a. Lateral rectus   d. Inferior rectus
   b. Medial rectus   e. Inferior oblique
   c. Superior rectus   f. Superior oblique
9. Lacrimal gland

B. Ear:

Locate and identify the following on models of the ear:

1. External ear
   a. Auricle
   b. External auditory canal
2. Middle ear (tympanic cavity)
   a. Tympanic membrane
   b. Ossicles
      1) Malleus
      2) Incus
      3) Stapes
   c. Pharyngotympanic tube (auditory tube, Eustachian tube)

3. Internal ear
    a. Semicircular canals – anterior, posterior, lateral
        1) Semicircular duct
        2) Ampulla (with crista ampullaris)
    b. Vestibule
        1) Utricle (with macula)
        2) Saccule (with macula)
        3) Oval window
        4) Round window

II. Histology:

A. Eye:

Identify the following on the **retina** slide:
1. Outer segment – rod and cone photoreceptor segments
2. Outer nuclear layer – nuclei of rods and cones
3. Outer plexiform layer – area of synapses between photoreceptors and bipolar cells
4. Inner nuclear layer – nuclei of bipolar cells
5. Inner plexiform layer – area of synapses between bipolar cells and ganglion cells
6. Nuclei of ganglion cells

B. Ear:

Identify the following on the **cochlea** model and slide:
1. Cochlear duct (scala media)
    a. Spiral organ (organ of Corti)
    b. Tectorial membrane
    c. Basilar membrane
2. Vestibular duct (scala vestibuli)
3. Vestibular membrane
4. Tympanic duct (scala tympani)
5. Spiral ganglion

CHAPTER 10

Digestive System

DIGESTIVE SYSTEM

I. Gross Anatomy:

Use charts, models, specimens, and cadavers to identify the following structures:

A. **Oral cavity**
   1. Hard palate
   2. Soft palate
      a. Uvula
   3. Tongue
      a. Lingual frenulum
      b. Papillae
         1) Vallate (circumvallate) papillae
   4. Extrinsic salivary glands and their ducts
      a. Parotid gland and duct
      b. Submandibular gland and ducts
      c. Sublingual gland and ducts

B. **Pharynx**
   1. Oropharynx
      a. Lingual tonsils
      b. Palatine tonsils
   2. Laryngopharynx
      a. Epiglottis

C. **Esophagus**

D. **Stomach**
   1. Greater curvature
   2. Lesser curvature
   3. Greater omentum (omenta, pl.)
   4. Lesser omentum
   5. Rugae
   6. Regions of the stomach
      a. Cardia (cardiac stomach)
      b. Fundus (fundic stomach)
      c. Body
      d. Pylorus (pyloric stomach)
   7. Pyloric sphincter

E. **Small intestine**
 1. Mesentery proper
 2. Duodenum
    a. Major duodenal papilla
    b. Plicae circulares *(this is the plural of plica circularis)*
 3. Jejunum
 4. Ileum
    a. Ileocecal valve

F. **Large Intestine**
 1. Diagnostic features:
    a. Haustra
    b. Teniae coli
    c. Epiploic appendages
 2. Regions:
    a. Cecum
       1) Appendix
    b. Colon
       1) Ascending colon
       2) Transverse colon
       3) Descending colon
       4) Sigmoid colon

G. **Liver**
 1. Lobes:
    a) Right lobe
    b) Left lobe
    c) Caudate lobe
    d) Quadrate lobe
 2. Common hepatic duct
 3. Common bile duct
 4. Common hepatic artery
 5. Hepatic portal vein
 6. Hepatic veins

H. **Pancreas**
    1. Main pancreatic duct (duct of Wirsung) (*view on model only*)
    2. Accessory pancreatic duct (duct of Santorini) (*view on model only*)

I. **Gallbladder**
    1. Cystic duct

II. Histology:

For all histological structures learn the specific tissues as well as functions.

Note: In the digestive system lab exam you will be asked to identify three structures covered in lectures prior to those on digestive system:
    1. lacteals – lymphatic capillaries (covered in lymphatic system)
    2. terminal ganglia – parasympathetic ganglia (covered in nervous system)
    3. GALT – gut associated lymphatic tissue (covered in lymphatic system)

A. **Tongue**
    1. Skeletal muscle tissue
    2. Papillae
        a. Filiform papillae
        b. Fungiform papillae
        c. Foliate papillae
        d. Vallate (circumvallate) papillae
            1) Intrinsic mucous glands
            2) Intrinsic serous glands
                a) Serous glands of von Ebner
        3. Taste buds (*associated with vallate, fungiform, and foliate papillae*)
            a) Supporting and gustatory cells (*you don't need to distinguish these cells*)

B. **Extrinsic salivary glands**
    1. Parotid gland
        a. Serous secretory units only
        b. Ducts
    2. Submandibular (submaxillary) gland
        a. Serous and mucous secretory units, mostly serous
        b. Ducts
    3. Sublingual gland
        a. Mucous secretory units, with scattered serous cells among them
        b. Ducts

C. **Esophagus**
    1. Tunica mucosa – non-keratinized stratified squamous epithelium, lamina propria, muscularis mucosae
    2. Tunica submucosa
        a. Esophageal glands
    3. Tunica muscularis (*transitions from skeletal muscle tissue to smooth muscle tissue*)
    4. Tunica adventitia

D. **Stomach**
1. Tunica mucosa – simple columnar epithelium, lamina propria, muscularis mucosae
   a. Pyloric stomach:
       1) Gastric pits and gastric glands, both lined with mucous cells
   b. Fundic stomach:
       1) Gastric pits with mucous cells
       2) Gastric glands with chief cells and parietal cells
2. Tunica submucosa
3. Tunica muscularis – 3 layers: oblique (*not usually visible*), circular, longitudinal
4. Tunica serosa

E. **Small intestine** (slides and histology models)
Be able to distinguish the three regions of small intestine: duodenum, jejunum, and ileum.

1. Tunica mucosa
   a. Simple columnar epithelium with goblet cells and microvilli (brush border), lamina propria, muscularis mucosae
   b. Villi
       1) Lacteals (*visible in model only)*
   c. Intestinal crypts (Crypts of Lieberkühn)
       1) Paneth cells
   d. Lymphatic nodules (GALT – gut associated lymphatic tissue) **may** be seen in lamina propria
2. Tunica submucosa – *this is the key to distinguishing the 3 regions*
   a. Duodenal (Brunner's) glands in duodenum
   b. Lymphatic nodules (GALT) **may** be seen in duodenum and jejunum
   c. Peyer's patches in ileum (GALT)
   d. Plicae circulares
3. Tunica muscularis – circular and longitudinal layers
4. Tunica serosa
5. Nervous tissue – *found in all parts of the GI tract but easier to identify here. Nervous tissue structures include two plexuses (networks) of nerve fibers and ganglia. You will be able to see only the ganglia in histological sections.*
   a. Submucosal (Meissner's) plexus
   b. Myenteric (Auerbach's) plexus
   c. Parasympathetic terminal ganglia (part of both plexuses)

F. **Large Intestine**
1. Tunica mucosa
   a. simple columnar epithelium with small microvilli and goblet cells, lamina propria, muscularis mucosae
   b. intestinal crypts
2. Tunica submucosa
3. Tunica muscularis – circular and longitudinal (teniae coli) layers
4. Tunica serosa

G. **Liver** (slide and histology models)
  1. Liver lobule
  2. Portal triad
        a. Hepatic ductule (bile ductule)
        b. Branch of common hepatic artery
        c. Branch of hepatic portal vein
  3. Hepatocytes
  4. Central vein
  5. Sinusoids
        a. Endothelial cells (*visible on model only*)
        b. Stellate reticuloendothelial (Kupffer) cells
  6. Bile canaliculi (*visible on model only*)

H. **Pancreas**
  1. Serous secretory units – *these are the exocrine cells*
  2. Pancreatic islets (islets of Langerhans) – *these are the endocrine cells*
  3. Ducts

MAIN FEATURES OF ALIMENTARY CANAL

REGION	DIAGNOSTIC FEATURES	TUNICA MUCOSA: ET, lamina propria, muscularis mucosae	TUNICA SUBMUCOSA	TUNICA MUSCULARIS	GLANDS
ESOPHAGUS	folded mucosa n-k stratified squamous ET adventitia, not serosa star-shaped lumen	non–keratinized, stratified squamous ET forms folds	esophageal glands	two layers: circular, longitudinal upper third is skeletal MT	esophageal glands in submucosa
STOMACH	rugae gastric pits oblique layer in muscularis no goblet cells no villi	simple columnar ET	forms large part of rugae	3 layers: oblique, circular, longitudinal oblique layer is non–continuous	gastric glands in lamina propria
DUODENUM	villi goblet cells plicae Brunner's glands	simple columnar ET with microvilli and goblet cells	forms inner part of plicae	two layers: circular, longitudinal	intestinal crypts in mucosa; Brunner's glands in submucosa
JEJUNUM	tall villi increasing goblet cells plicae	simple columnar ET with microvilli and goblet cells	forms inner part of plicae	two layers: circular, longitudinal	intestinal crypts in mucosa
ILEUM	shorter villi increasing goblet cells few plicae Peyer's patches	simple columnar ET with microvilli and goblet cells ; Peyer's patches in lamina propria	Peyer's patches extend into submucosa	two layers: circular, longitudinal	intestinal crypts in mucosa
LARGE INTESTINE	no villi many goblet cells teniae coli	simple columnar ET; large lymphatic follicles (GALT) in lamina propria	lymphatic follicles (GALT) may extend into submucosa	two layers, circular, longitudinal (forms teniae coli)	intestinal crypts in mucosa

Santa Rosa Junior College–Biological Sciences Department

CHAPTER 11

Respiratory System

RESPIRATORY SYSTEM

I. Gross Anatomy: Use charts, models, specimens, and cadavers to identify the following structures:

A. **Nose**
   1. External nares (plural)
   2. Nasal conchae (turbinates)
      a. Superior nasal conchae
      b. Middle nasal conchae
      c. Inferior nasal conchae
   3. Hard palate
   4. Soft palate
   5. Paranasal sinuses (frontal, maxillary, sphenoidal, ethmoidal)

B. **Pharynx**
   1. Nasopharynx
      a. Pharyngeal (adenoid) tonsil
      b. Pharyngotympanic tube (auditory, Eustachian) tube
   2. Oropharynx
      a. Palatine tonsils
      b. Lingual tonsils
      c. Uvula
   3. Laryngopharynx

C. **Larynx**
   1. Cartilages of the larynx:
      a. Thyroid cartilage   d. Arytenoid cartilages
      b. Cricoid cartilage   e. Cuneiform cartilages
      c. Epiglottis   f. Corniculate cartilages
   2. Ventricular (vestibular) folds (false vocal cords)
   3. Glottis
      a. Vocal folds (true vocal cords)
      b. Rima glottidis

D. **Hyoid**

E. **Trachea**
   1. Tracheal rings
   2. Trachealis muscle

F. **Respiratory tree**
   1. Primary bronchi (*serve lungs, have cartilage rings*)
   2. Secondary bronchi (*serve lobes of the lungs, have cartilage plates*)
   3. Tertiary bronchi (*serve bronchopulmonary segments, have cartilage plates*)

G. **Pulmonary arteries and veins**

H. **Lungs**
1. Visceral pleura
2. Lobes – 3 in right lung, 2 in left lung
3. Cardiac notch
4. Hilus
5. Root

II. **Histology:**

A. **Trachea**
1. Tunica mucosa–
   pseudostratified ciliated columnar epithelium with goblet cells
2. Tunica submucosa – thick, with many tracheal glands *(only visible on some slides)*
3. Tracheal rings – hyaline cartilage
4. Trachealis muscle – smooth muscle tissue

B. **Lung**
1. Bronchi *(secondary and tertiary bronchi have cartilage plates)*
2. Bronchioles *(have smooth muscle, but no cartilage in wall)*
3. Alveolar ducts
4. Alveolar sacs
   a. Alveoli
      1) Type I alveolar cells (simple squamous epithelium)
      2) Alveolar macrophages (dust cells)
      3) Type II alveolar cells (septal cells)

COMPARABLE HISTOLOGICAL SUMMARY OF THE MAJOR SUBDIVISIONS OF THE TUBULAR ORGANS OF THE RESPIRATORY SYSTEM

Characteristics	Trachea	1° Bronchus	2° Bronchus	3° Bronchus	Terminal Bronchiole	Respiratory Bronchiole	Alveolar Duct
epithelium	pseudostratified ciliated columnar with goblet cells	pseudostratified ciliated columnar with goblet cells	pseudostratified ciliated columnar with goblet cells	pseudostratified ciliated columnar with goblet cells	simple columnar	simple cuboidal	simple squamous
cartilage	rings	rings	plates	plates	absent	absent	absent
smooth muscle	present (trachealis muscle)	present	present	present	abundant	abundant	few fibers present
seromucous glands	present (tracheal glands)	present	present	present	absent	absent	absent
alveoli	absent	absent	absent	absent	absent	present	abundant

CHAPTER 12

Urinary System

Urinary System

I. Gross Anatomy:

Use charts, models, specimens, and cadavers to identify the following structures:

A. Kidneys
1. Cortex
   a. Cortical arches
   b. Renal columns
2. Medulla
   a. Renal pyramids
      1) Renal papillae
3. Renal sinus
   a. Minor calyces
   b. Major calyces
   c. Renal pelvis
4. Blood supply at renal hilus
   a. Renal artery
   b. Renal vein

B. Ureters

C. Urinary bladder
1. Rugae
2. Trigone
3. Detrusor muscle

D. Urethra – male and female

E. Adrenal gland

II. Histology:

A. **Kidney**

Identify the following using slides and models;
structures marked * are only visible on models:

1. Cortex
   a. Renal corpuscle
      1) Glomerular capsule (Bowman's capsule)
         a) Parietal layer – simple squamous epithelium
         b) Visceral layer – podocytes, filtration slits*
         c) Capsular space
      2) Glomerulus
   b. Proximal convoluted tubules
   c. Distal convoluted tubules
   d. Juxtaglomerular apparatus
      1) Macula densa (*modified cells of the DCT*)
      2) Juxtaglomerular cells* (*modified cells of the afferent arteriole*)
2. Medulla
   a. Descending limb of the nephron loop (loop of Henle)
   b. Ascending limb of the nephron loop (loop of Henle)
   c. Collecting duct
3. Blood supply
   a. Afferent arteriole*
   b. Glomerulus*
   c. Efferent arteriole*
   d. Peritubular capillaries*
      1. Vasa recta*

B. **Ureter**

1. Tunica mucosa – transitional epithelium
2. Tunica muscularis
3. Adventitia

C. **Urinary Bladder**

1. Tunica mucosa – transitional epithelium
2. Detrusor muscle – smooth muscle tissue

CHAPTER 13

Reproductive System

Reproductive System

I. Gross Anatomy:

Use charts, models, specimens, and cadavers to identify the following structures:

A. Male Reproductive System
1. Scrotum
    a. Dartos muscle
2. Testis *(pl. testes)*
    a. Seminiferous tubules
    b. Tunica albuginea
    c. Tunica vaginalis
3. Epididymis
4. Ductus (vas) deferens
5. Ejaculatory duct
6. Accessory glands
    a. Seminal vesicles
    b. Prostate gland
    c. Bulbourethral (Cowper's) gland (*not visible on prosections*)
7. Urethra
    a. Prostatic urethra
    b. Membranous urethra
    c. Spongy (penile) urethra
8. Penis
    a. Glans penis
        1. Corona
    b. Prepuce (foreskin)
    c. Erectile tissue of body of penis:
        1) Corpora cavernosa (two)
        2) Corpus spongiosum (one)
    d. Root of penis
        1) Crura of penis (two)
        2) Bulb of penis (one)
9. Inguinal canal
10. Spermatic cord
    a. Testicular nerve
    b. Testicular artery
    c. Testicular vein (pampiniform plexus)
    e. Ductus (vas) deferens
    d. Cremaster muscle

B. **Female Reproductive System**

1. Ovary
   a. Ovarian ligament
   b. Suspensory ligament
2. Fallopian Tube
   a. Fimbriae
3. Uterus
   a. Fundus
   b. Body
   c. Cervix
   d. Broad ligament
   e. Round ligament
4. Vagina
   a. Fornix
5. Vulva (*external genitalia*)
   a. Mons pubis
   b. Labia majora
   c. Labia minora
   d. Vestibule
      1) External urethral orifice
      2) External vaginal orifice
   e. Clitoris
      1) Glans
      2) Body of the clitoris (*view on model only*)
      3) Crus of the clitoris (*view on model only*)
      4) Bulb of the vestibule (*view on model only*)
6. Breast
   a. Mammary glands
   b. Nipple
   c. Areola

II. Histology

A. Male Reproductive System

1. **Testis**:
   a. Seminiferous tubules
      1) Developing spermatocytes
      2) Spermatozoa (*in lumen)*
      3) Sustentacular (Sertoli) cells
   b. Interstitial (Leydig) cells
2. **Epididymis**:
   a. Ductus epididymis
      1) Pseudostratified columnar ET with stereocilia (long microvilli)
3. **Penis**:
   a. Corpora cavernosa (*may be single in primate slides*)
   b. Corpus spongiosum
      1) Spongy (penile) urethra
   c. Tunica albuginea

B. Female Reproductive System

1. **Ovary** (slide and histology model):
   a. Germinal epithelium
   b. Stroma
   c. Ovarian follicles
      1) Primordial follicle
         a) Primary oocyte
      2) Primary follicle
         a) Primary oocyte
      3) Secondary follicle
         a) Primary oocyte
         a) Antrum (*small*)
      4) Mature (Graafian) follicle
         a) Secondary oocyte
         b) Antrum (*large*)
         c) Zona pellucida
         d) Corona radiata
         e) Granulosa cells
         f) Thecal cells
      5) Corpus luteum

2. **Uterus** (*slide from secretory phase of the uterine cycle*):
   a. Myometrium
   b. Endometrium
      1) Stratum functionalis
      2) Stratum basalis
      3) Endometrial (uterine) glands

CHAPTER 14

Dissections

Dissections

Introduction:

In the laboratory periods associated with the muscular system you will begin the dissection of the human cadavers. As a rule you will work on your dissection with a partner, but occasionally students may work alone. If you work as a team, each student is expected to do their share of the dissection and the cleanup. Each dissector will expose, clean, and separate muscles from a specific body region and make these structures available to your classmates to study. They in turn will make a different region available to you. Dissections are worth 50 points. Scalpels, forceps, probes, etc. will be provided. Gloves and lab coats need to be supplied by students.

Proper Care of Cadaver:

When a student is assigned to a cadaver, they assume responsibility for its proper care. You will find the cadaver already preserved or embalmed, and the whole body kept moist by maintaining adequate wrappings. Uncover only those body parts on which you dissect. Inspect every part periodically, and renew and moisten wrappings as occasion demands. No part must ever be left exposed to the air needlessly. Special attention must be given to the face, hands, feet and external genitalia. Once a part is allowed to become dry and hard it can never be fully restored, and its proper dissection is impossible. Every time you work in the cadaver lab you will be expected to leave the lab as clean or cleaner than you found it. Instruments should be cleaned and put away, waste materials put in appropriate receptacles, the cadavers sprayed with Carosafe and covered as demonstrated in lab, and the floor mopped in the area in which you worked.

Scheduling:

A student's dissection must be completed during the lab period in which it is assigned. Because lab time is limited, it is beneficial to be as prepared as possible before you even begin. Research your assigned area beforehand. You will also need to make arrangements to make up the regularly scheduled lab that you missed while completing your dissection. You instructor will provide details about this.

Purpose of dissection:

There is no substitute for a three-dimensional approach to the structures of the body. Observe and palpate the topographic relations of various structures to each other. Feel the texture of blood vessels, nerves, and various tissues. Test the rigidity of bones and the strength of ligaments. Explore and appreciate the three dimensions of anatomical structures.

Characteristics of a Quality Dissection:

Care should be given to each dissection. Following the procedures outlined in lab, each structure on your dissection list should be isolated, cleaned, and separated from surrounding tissues. Dissected structures should be intact, uncut, and untorn unless a particular cut has been indicated as necessary to expose underlying structures. Please be careful not to damage tissues/structures that may be part of another student's dissection. The use of cadavers is a privilege and an extremely valuable resource to the class as a whole. Think before you cut. In the final analysis, the muscle mass and isolated structures should appear as they do in the textbooks.

Efficiency:

Time in lab is valuable. the following suggestions will help you to learn as much as possible in the limited time we have:

A. Acquire a theoretical concept of the area under investigation before you attempt to dissect it. You do not "dig around" and happen to find "something interesting." You must deliberately search for certain structures.

B. Use a good anatomical atlas (available in lab).

C. Always palpate bony landmarks since they are keys in your search for related soft structures.

D. Use your time wisely. To spend an hour tracing the terminal twigs of a cutaneous nerve, when the general skin area supplied by the nerve is obvious, is spending an hour for little gain. To spend 3 minutes to define the exact fiber direction of a ligament is to spend 3 minutes for great gain; you will understand why and how that ligament restrains or prevents certain movements of bony structures.

E. Demonstrate the essential features of a given anatomical region with clarity. Remove fat, areolar connective tissue, and smaller veins. If a clear display of arteries is obtained, the general arrangement of the companion veins will be obvious.

Dissecting Techniques:

Keep in mind that a variable amount of subcutaneous fat lies immediately deep to the skin. In those cadavers nearly devoid of fat, one needs to exercise special care in order not to go too deep. If, during removal of skin, you see brownish muscular fibers shining through the filmy deep fascia, you are too deep. Always remember to put tension on the skin as it is being removed, to keep the sharp scapel blade directed against it, and to remove as much fat as possible with the skin. In this manner, you will work faster and encounter fewer difficulties.

How will my dissection be graded?

Evaluation Criteria:

- Preparation for dissection clearly demonstrated

- Dissectors work carfully and thoughtfully for entire dissection period

- Structures on list clean and easily observable (separated from surrounding structures)

- Muscles should be dissected from origin to insertion, except as noted

- Structures on list intact, uncut, and untorn

- Surrounding structures not carelessly cut/torn

- Overall quality of dissection

MUSCULAR SYSTEM

Day 1 Dissection

Lab Section: Green   Purple   Blue   Yellow   Salmon   Pink

Cadaver #: _____   Students: 1._____

Sex: _____   2._____

Side of body: _____

Assignment date: _____

Head, Neck, Back

_____ Latissimus dorsi

_____ Trapezius

_____ Rhomboid major

_____ Rhomboid minor

_____ Levator scapulae

_____ *Preparation*

_____ *Overall quality*

Notes on Muscle Dissection 1

Head, Neck & Back:

- Trapezius must be cut and reflected to reveal the rhomboid muscles. Separate the rhomboids from the trapezius by hand and place a scalpel handle under the trapezius while making the cut.

- Do not dissect latissimus dorsi to its insertion. Instead, stop just beyond the axillary line.

- Be careful not to damage the sternocleidomastoid and external jugular veins. These are anterolateral and should be left undisturbed.

MUSCULAR SYSTEM

Day 1 Dissection

Lab Section: Green   Purple   Blue   Yellow   Salmon   Pink

Cadaver #: _____   Students: 1._____

Sex: _____   2._____

Side of body: _____

Assignment date: _____

Calf

_____ Gastrocnemius

_____ Soleus

_____ Plantaris

_____ Popliteus

_____ Fibularis longus

_____ Common fibular nerve

_____ Tibial nerve

_____ Popliteal artery

_____ Popliteal vein

_____ *Preparation*

_____ *Overall quality*

Notes on Muscle Dissection 1

Calf:

- Gastrocnemius must be cut and reflected to reveal the soleus, plantaris, and popliteus.

- Before making the reflection cut, separate the gastrocnemius from the deep muscles. Locate and separate the plantaris tendon so that it is not cut during reflection.

- Place a scalpel handle under the gastrocnemius and over the soleus and plantaris tendon before cutting.

- Vessels and nerves should be separated and cleaned to approximately 4 inches above the knee joint including the nerve branch.

- Fascia must be removed from the popliteus.

MUSCULAR SYSTEM

Day 2 Dissection

Lab Section: Green   Purple   Blue   Yellow   Salmon   Pink

Cadaver #: _____   Students: 1._____

Sex: _____   2._____

Side of body: _____

Assignment date: _____

Posterior Upper Limb

_____ Supraspinatus

_____ Infraspinatus

_____ Deltoid (posterior part)

_____ Triceps brachii (3 heads)

_____ Lateral head

_____ Medial head

_____ Long head

_____ Teres major

_____ Teres minor

_____ *Preparation*

_____ *Overall quality*

Notes on Muscle Dissection 2

Superior Dissection:

- You don't have to dissect to the insertion on the following muscles: supraspinatus, infraspinatus, teres minor and the long head of the triceps brachii.

- Remove the fascia from the supraspinatus and infraspinatus.

- Be careful not to cut brachial arteries, veins, or nerve branches of the brachial plexus.

MUSCULAR SYSTEM

Day 2 Dissection

Lab Section: Green   Purple   Blue   Yellow   Salmon   Pink

Cadaver #: _____   Students: 1._____

Sex: _____   2._____

Side of body: _____

Assignment date: _____

Gluteal Region and Posterior Thigh

_____ Biceps femoris, long head

_____ Biceps femoris, short head

_____ Semitendinosus

_____ Semimembranosus

_____ Gluteus maximus

_____ Gluteus medius

_____ *Preparation*

_____ *Overall quality*

Notes on Muscle Dissection 2

Inferior Dissection:

- Gluteus maximus must be cut and reflected to reveal the gluteus medius.

- Be careful not to cut the sciatic nerve.

- Leave the iliotibial band intact.

MUSCULAR SYSTEM

Day 3 Dissection

Lab Section: Green   Purple   Blue   Yellow   Salmon   Pink

Cadaver #: _____   Students: 1._____

Sex: _____   2._____

Side of body: _____

Assignment date: _____

Thorax and Abdomen

_____ External intercostals

_____ Internal intercostals

_____ External obliques

_____ Internal obliques

_____ Transverse abdominis

_____ Rectus abdominis

_____ Serratus anterior

_____ *Preparation*

_____ *Overall quality*

Notes on Muscle Dissection 3

Superior Dissection:

- Reveal the external and internal intercostals by cutting a window through the external obliques.

- External intercostals project only partially onto the anterior surface. It is only required that you expose the last few inches of a few intercostal muscles.

- Reveal the internal obliques and transverse abdominis by cutting a window or flap through the external obliques such that the muscles are exposed and layering can be visualized.

MUSCULAR SYSTEM

Day 3 Dissection

Lab Section: Green   Purple   Blue   Yellow   Salmon   Pink

Cadaver #: _____   Students: 1._____

Sex: _____   2._____

Side of body: _____

Assignment date: _____

Anterior and Medial Thigh

_____ Rectus femoris

_____ Vastus lateralis

_____ Vastus medialis

_____ Vastus intermedius

_____ Sartorius

_____ Gracilis

_____ Adductor longus

_____ Adductor brevis

_____ Adductor magnus

_____ Saphenous vein

_____ *Preparation*

_____ *Overall quality*

Notes on Muscle Dissection 3

Inferior Dissection:

- Be sure to dissect to the origin of all muscles except adductors and gracilis. For these muscles get as close as possible without damaging the femoral nerves and vessels.

- The saphenous vein is very superficial and should be found before any adipose tissue removal on the medial thigh.

- Be careful not to sever the iliotibial band.

MUSCULAR SYSTEM

Day 4 Dissection

Lab Section: Green   Purple   Blue   Yellow   Salmon   Pink

Cadaver #: _____   Students: 1._____

Sex: _____   2._____

Side of body: _____

Assignment date: _____

Anterior Upper Limb

_____ Pectoralis major

_____ Pectoralis minor

_____ Biceps brachii

_____ Long head

_____ Short head

_____ Brachialis

_____ Coracobrachialis

_____ Cephalic vein

_____ External jugular vein

_____ Sternocleidomastoid

_____ Deltoid (anterior portion)

_____ *Preparation*

_____ *Overall quality*

Notes on Muscle Dissection 4

Superior Dissection:

- Pectoralis major must be cut and reflected to reveal pectoralis minor. Separate pectoralis major from the underlying muscle before placing a scalpel handle beneath to guard against cutting too deep.

- Do not separate sternocleidomastoid from underlying muscles.

- The external jugular and cephalic veins are very superficial. Skinning in this region must be performed very superficially.

MUSCULAR SYSTEM

Day 4 Dissection

Lab Section: Green   Purple   Blue   Yellow   Salmon   Pink

Cadaver #: _____   Students: 1._____

Sex: _____   2._____

Side of body: _____

Assignment date: _____

Knee Joint and Tibialis Anterior

_____ Lateral meniscus

_____ Medial meniscus

_____ Anterior cruciate ligament

_____ Transverse ligament

_____ Fibular collateral ligament

_____ Tibial collateral ligament

_____ Tibialis anterior

_____ *Preparation*

_____ *Overall quality*

Notes on Muscle Dissection 4

Inferior Dissection:

- Cut and reflect rectus femoris.

- Cut along the patellar tendon just on either side of the patella.

- When separating patella from knee joint, separate the connective tissue from the posterior side of the patella leaving the connective tissue inside the capsule.

- Carefully clean the capsule while searching for the transverse ligament.

- Fibular collateral ligament and tibial collateral ligament can be located by feeling for them with your finger while gently moving the joint in the frontal plane.

Coelom and Viscera

Day 1 Dissection

Lab Section: Green   Purple   Blue   Yellow   Salmon   Pink

Cadaver #: _____   Students: 1._____

Sex: _____   2._____

Side of body: _____

Assignment date: _____

Thoracic Cavity and Mediastinum

_____ Trachea (right side dissectors)

_____ Thyroid gland

_____ Superior vena cava (right side dissectors)

_____ Brachiocephalic trunk (artery) (right side dissectors)

_____ Thyro–cervical artery (trunk)

_____ Common carotid artery

_____ Internal and external carotid arteries

_____ Brachiocephalic vein

_____ Vertebral artery and vein

_____ Ligamentum arteriosum (left side dissectors)

_____ Internal and external jugular veins

_____ Subclavian artery and vein

_____ Cephalic vein

_____ Pulmonary artery and vein – from the heart to the lungs

_____ Azygos vein (right side dissectors)

_____ Aorta – ascending, arch of, descending (left side dissectors)

_____ Vagus nerve

_____ Phrenic nerve

_____ Recurrent laryngeal nerve – from the branching of vagus nerve

_____ *Preparation*

_____ *Overall quality*

Coelom and Viscera

Day 2 Dissection

Lab Section: Green   Purple   Blue   Yellow   Salmon   Pink

Cadaver #: _____   Students: 1._____

Sex: _____   2._____

Side of body: _____

Assignment date: _____

Lower Peritoneal Cavity–Renal Trunk to Pelvic Wall Exit

_____ Large intestines

_____ Small intestines

_____ Cecum (right side)

_____ Appendix (right side)

_____ Urinary bladder

_____ Ureters

_____ Gonadal artery and vein

_____ Abdominal aorta

_____ Inferior mesenteric vein (left side)

_____ Inferior mesenteric artery (right side)

_____ Common iliac artery and vein

_____ Internal iliac artery and vein

_____ External iliac artery and vein

_____ Round ligament (female)

_____ Vas deferens (male)

_____ *Preparation*

_____ *Overall quality*

Coelom and Viscera

Day 3 Dissection

Lab Section:　　Green　　Purple　　Blue　　Yellow　　Salmon　　Pink

Cadaver #: _____　　Students: 1._____

Sex: _____　　　　　　　　2._____

Side of body: _____

Assignment date: _____

Upper Peritoneal Cavity

_____ Greater and lesser omenta – observed (student may remove to access deeper structures)

_____ Liver and gallbladder (right side)

_____ Stomach (left side)

_____ Spleen (left side)

_____ Pancreas (left side)

_____ Kidney

_____ Adrenal gland

_____ Descending aorta to renal trunk

_____ Celiac trunk – isolate celiac trunk to:

　　　　gastric artery (left)
　　　　common hepatic artery (right)
　　　　splenic artery (left)
　　　　　　and follow them to the organs they nourish

_____ Superior mesenteric artery (right side) – follow a short distance into mesentery

_____ Superior mesenteric vein (right side)

_____ Hepatic portal vein (right side)

_____ Splenic vein (left side)

_____ Inferior mesenteric vein (left side)

_____ Gastric vein (left side)

_____ Inferior vena cava

_____ Common bile duct (right side)

_____ Cystic duct (right side)

_____ Common hepatic duct (right side)

_____ Renal artery and vein

_____ Suprarenal artery and vein

_____ *Preparation*

_____ *Overall quality*

Coelom and Viscera

Day 4 Dissection

Lab Section: Green   Purple   Blue   Yellow   Salmon   Pink

Cadaver #: _____   Students: 1._____

Sex: _____   2._____

Side of body: _____

Assignment date: _____

Inguinal Ligament & Canal, Femoral Triangle, Reproductive Structures, Brachial Plexus

_____   Inguinal ligament

_____   Inguinal canal and its contents

_____   Femoral nerve, artery, vein, and venous valve

_____   Saphenous vein

_____   Female reproductive structures

         _____   Uterus

         _____   Ovaries

         _____   Fallopian tubes

_____   Male reproductive structures

         _____   Spermatic cord:

         _____   Testicular artery and vein

         _____   Ductus deferens

         _____   Tunica vaginalis

         _____   Tunica albuginea

         _____   Epididymis

         _____   Testis

_____   Brachial Plexus

         _____   Radial nerve

         _____   Median nerve

         _____   Ulnar nerve

_____   *Preparation*

_____   *Overall quality*

FOREARM DISSECTION

Cadaver #: _____   Students: 1._____

Sex: _____   2._____

Side of body: Right or Left

Assignment date: _____

Forearm

- Palmaris longus

- Flexor carpi radialis

- Flexor carpi ulnaris

- Extensor carpi radialis longus

- Extensor carpi radialis brevis

- Extensor carpi ulnaris

- Extensor digitorum

- Brachioradialis

- Retinaculum

HEART DISSECTION

Cadaver #: _____   Students: 1._____

Sex: _____   2._____

Side of body: Right or Left

Assignment date: _____

Heart

Cut the following to free the heart from the cadaver:

- Aorta with attached major vessels, approx. 1 inch of each

- Superior vena cava with attached major vessels, approx. 1 inch of each

- Inferior vena cava

- Pulmonary trunk with L & R pulmonary arteries

- All 4 pulmonary veins

- Pericardial sac, superior to the diaphragm

Remove the heart, and then do the following:

- Clean up the outside to see coronary arteries, veins, and sinus; remove all extraneous tissues.

- Make a window into right or left atrium and its corresponding ventricle to expose the atrioventricular valve and semilunar valve.

- Clean coagulated blood from inside to clearly reveal the papillary muscles and chordae tendineae.

Appendix: Word Roots, Prefixes, Suffixes, and Combining Forms

PREFIXES AND COMBINING FORMS

a-, an- *absence or lack* acardia, lack of a heart: anaerobic, in the absence of oxygen

ab- *departing from: away from* abnormal, departing from normal

acou- *hearing* acoustics, the science of sound

acr-, acro- *extreme or extremity; peak* acrodermatitis, inflammation of the skin of the extremities

ad- *to or toward* adorbital, toward the orbit

aden-, adeno- *gland* adeniform, resembling a gland in shape

adren- *toward the kidney* adrenal gland, adjacent to the kidney

aero- *air* aerobic respiration; oxygen-requiring metabolism

af- *toward* afferent neurons, which carry impulses to the central nervous system

agon- *contest* agonistic and antagonistic muscles, which oppose each other

alb- *white* corpus albicans of the ovary, a white scar tissue

aliment- *nourish* alimentary canal or digestive tract

allel- *of one another* alleles, alternative expressions of a gene

amphi- *on both sides; of both kinds* amphibian, an organism capable of living in water and on land

ana- *apart, up; again* anaphase of mitosis, when the chromosomes separate

anastomos- *come together* arteriovenous anastomosis, a connection between an artery and a vein

aneurysm *a widening* aortic aneurism, a weak spot that causes enlargement of the blood vessel

angi- *vessel* angiitis. inflammation of a lymph vessel or blood vessel

angin- *choked* angina pectoris, a choked feeling in the chest due to dysfunction of the heart

ant-, anti- *opposed to; preventing or inhibiting* anticoagulant, a substance that prevents blood coagulation

ante- *preceding, before* antecubital, in front of the elbow

aort- *great artery* aorta

ap-, api- *tip, extremity* apex of the heart

append- *hang to* appendicular skeleton

aqua-, aque- *water* aqueous solutions

arbor- *tree* arbor vitae of the cerebellum, the tree-like pattern of white matter

areola- *open space* areolar connective tissue, a loose connective tissue.

arrect- *upright* arrector pili muscles of the skin, which make the hairs stand erect

arthr-, arthro- *joint* arthropathy, any joint disease

artic-. *joint* articular surfaces of bones, the points of connection

atri- *vestibule* atria, upper chambers of the heart

auscult- *listen* auscultatory method for measuring blood pressure

aut- auto- *self* autogenous, self-generated

ax- axi- axo- *axis, axle* axial skeleton, axis of vertebral column

azyg- *unpaired* azygous vein, an unpaired vessel

baro- *pressure* baroreceptors for monitoring blood pressure

basal- *base* basal lamina of epithelial basement membrane

bi- *two* bicuspid, having two cusps

bili- *bile* bilirubin, a bile pigment

bio- *life* biology, the study of life and living organisms

blast- *bud or germ* blastocyte, undifferentiated embryonic cell

brachi- *arm* brachial plexus of peripheral nervous system supplies the arm

brady- *slow* bradycardia, abnormally slow heart rate

brev- *short* peroneus brevis, a short leg muscle

broncha- *bronchus* bronchospasm, spasmodic contraction of bronchial muscle

bucco- *cheek* buccolabial, pertaining to the cheek and lip

calor- *heat* calories, a measure of energy

capill- *hair* blood and lymph capillaries

caput- *head* decapitate, remove the head

carcin- *cancer* carcinogen, a cancer-causing agent

cardi, cardio- *heart* cardiotoxic, harmful to the heart

carneo- *flesh* trabeculae carneae, ridges of muscle in the ventricles of the heart

carat 1) *carrot,* 2) *stupor* carotene, an orange pigment; 2) carotid arteries in the neck, blockage causes fainting

cata- *down* catabolism, chemical breakdown

caud- *tail* caudal (directional term)

cec- *blind* cecum of large intestine, a blind-ended pouch

cele- *abdominal* celiac artery, in the abdomen

cephal- *head* cephalometer, an instrument for measuring the head

cerebro- *brain, especially the cerebrum* cerebrospinal, pertaining to the brain and spinal cord

cervic-, cervix- *neck* cervix of the uterus

chiasm- *crossing* optic chiasma, where optic nerves cross

chole- *bile* cholesterol; cholecystokinin, a bile-secreting hormone

chondr- *cartilage* chondrogenic, giving rise to cartilage

chrom- *colored* chromosome, so named because they stain darkly

cili- *small hair* ciliated epithelium

circum- *around* circumnuclear, surrounding the nucleus

clavic- *key* clavicle, a "skeleton key"

co-, con- *together* concentric, common center, together in the center

coccy- *cuckoo* coccyx, which is beak-shaped

cochlea- *snail shell* the cochlea of the inner ear, which is coiled like a snail shell

coel- *hallow* coelom, the ventral body cavity

commis- *united* gray commissure of the spinal cord connects the two columns of gray matter

concha- *shell* nasal conchae, coiled shelves of bone in the nasal cavity

contra- *against* contraceptive, agent preventing conception

corn-, cornu- *horn* stratum corneum, outer layer of the skin composed of (horny) cells

corona- *crown* coronal suture of the skull

corp- *body* corpse. corpus luteum, hormone-secreting body in the ovary

cort- *bark* cortex, the outer layer of the brain, kidney, adrenal glands, lymph nodes, ovaries

cost- *rib* intercostal, between the ribs

crani- *skull* craniotomy, a skull operation

crypt- *hidden* cyptomenorrhea, a condition in which menstrual symptoms are experienced but no external loss of blood occurs

cusp- *pointed* bicuspid, tricuspid valves of the heart

cutic- *skin* cuticle of the nail

cyan- *blue* cyanosis, blue color of the skin due to lack of oxygen

cyst- *sac, bladder* cystitis, inflammation of the urinary bladder

cyt- *cell* cytology, the study of cells

de- *undoing. reversal, loss, removal* deactivation, becoming inactive

decid- *falling off* deciduous (milk) teeth

delta- *triangular* deltoid muscle, roughly triangular in shape

den-, dent- *tooth* dentin of the tooth

dendr- *tree, branch* dendrites, telodendria, both branches of a neuron

derm- *skin* dermis, deep layer of the skin

desm- *bond* desmosome, which binds adjacent epithelial cells

di- *twice, double* dimorphism, having two forms

dia- *through, between* diaphragm, the wall through or between two areas

dialys- *separate, break apart* kidney dialysis, in which waste products are removed from the blood

diastol- *stand apart* cardiac diastole, between successive contractions of the heart

diure- *urinate* diuretic, a drug that increases urine output

dors- *the back* dorsal; dorsum; dorsiflexion

duc-, duct- *lead, draw* ductus deferens which carries sperm from the epididymus into the urethra during ejaculation

dura-*hard* dura mater, tough outer meninx

dys- *difficult, faulty,painful* dyspepsia, disturbed digestion

ec-, ex-, ecto- *out, outside, awayfrom* excrete, to remove materials from the body

ectop- *displaced* ectopic pregnancy, ectopic focus for initiation of heart contraction

edem- *swelling* edema, accumulation of water in body tissues

ef- *away* efferent nerve fibers, which carry impulses away from the central nervous system

ejac- *to shoot forth* ejaculation of semen

embol- *wedge* embolus, an obstructive object traveling in the bloodstream

en-, em- *in, inside* encysted, enclosed in a cyst or capsule

enceph.- *brain* encephalitis, inflammation of the brain

endo- *within, inner* endocytosis, taking particles into a cell

entero- *intestine* enterologist, one who specializes in the study of intestinal disorders

epi- *over, above* epidermis, outer layer of skin

erythr- *red* erythema, redness of the skin; erythrocyte, red blood cell

eso- *within* esophagus

eu- *well* euesthesia, a normal state of the senses

excret- *separate* excretory system

exo- *outside, outer layer* exophthalmos, an abnormal protrusion of the eye from the orbit

extra- *outside, beyond* extracellular, outside the body cells of an organism.

extrins- *from the outside* extrinsic regulation of the heart

fasci-, fascia- *bundle, band* superficial and deep fascia

fenestr- *window* fenestrae of the inner ear; fenestrated capillaries

ferr- *iron* transferrin, ferritin, both iron-storage proteins

flagell- *whip* flagellum, the tail of a sperm cell

flat- *blow, blown* flatulence

folli- *bag, bellows* hair follicle

fontan- *fountain* fontanels of the fetal skull

foram- *opening-* foramen magnum of the skull

foss- *ditch* fossa ovalis of the heart: mandibular fossa of the skull

gam- gamet- *married, spouse* gametes, the sex cells

gangli- *swelling, or knot* dorsal root ganglia of the spinal nerves

gastr- *stomach* gastrin, a hormone that influences gastric acid secretion

gene- *beginning; origin* genetics

germin- *grow* germinal epithelium of the gonads

gem-, geront- *old man* gerontology, the study of aging

gest- *carried* gestation, the period from conception to birth

glauc- *gray* glaucoma, which causes gradual blindness

glom- *ball* glomeruli, clusters of capillaries in the kidneys

glosso- *tongue* glossopathy, any disease of the tongue

gluco-, glyco- gluconeogenesis. the production of glucose from noncarbohydrate molecules

glute- *buttock* gluteus maximus. the largest muscle of the buttock

gnost- *knowing* the gnostic sense, a sense of awareness of self

gompho- *nail* gomphosis, the term applied to the joint between tooth and jaw

gon- gono- *seed, offspring* gonads, the sex organs

gust- *taste* gustatory sense, the sense of taste

hapt- *fasten, grasp* hapten, a partial antigen

hema-, hemato., hemo- *blood* hematocyst. a cyst containing blood

hemi- *half* hemiglossal, pertaining to one-half of the tongue

hepat- *liver* hepatitis, inflammation of the liver

hetero- *different or other* heterosexuality, sexual desire for a person of the opposite sex

hiat *gap* the hiatus of the diaphragm, the opening through which the esophagus passes

hippo- *horse* hippocampus of the brain, shaped like a seahorse .

hirsut- *hairy* hirsutism, excessive body hair

hist- *tissue* histology, the study of tissues

holo- *whole* holocrine glands, whose secretions are whole cells

hom-, homo- *same* homeoplasia, formation of tissue similar to normal tissue; homocentric, having the same center

hormon- *to excite* hormones

humor- *a fluid* humoral immunity, which involves antibodies circulating in the blood

hyal- *clear* hyaline cartilage, which has no visible fibers

hydr-, hydro- *water* dehydration, loss of body water

hyper- *excess* hypertension, excessive tension

hypno- *sleep* hypnosis, a sleeplike state

hypo- *below, deficient* hypodermic, beneath the skin; hypokalemia, deficiency of potassium

hyster-, hystero- *uterus or womb* hysterectomy, removal of the uterus; hysterodynia. pain in the womb

ile- *intestine* ileum, the last portion of the small intestine

im- *not* impermeable, not permitting passage, not permeable

inter- *between* intercellular, between the cells

intercal- *insert* intercalated disc, the end membranes between adjacent cardiac muscle

intra- *within, inside* intracellular, inside the cell

iso- *equal, same* isothermal, equal or same temperature

jugul- *throat* jugular veins, prominent vessels in the neck

juxta- *near, close to* juxtaglomerular apparatus, a cell cluster next to the glomeruli in the kidneys

karyo- *kernal nucleus* karyotype, the assemblage of the nuclear chromosomes

kera- *horn* keratin, the water-repellent protein of the skin

kilo- *thousand* kilocalories, equivalent to one thousand calories

kin- kines- *move* kinetic energy, the energy of motion

lahi- lahri- *lip* labial frenulum. the membrane which joins the lip to the gum

lact- *milk* lactose milk sugar

lacun- *space, cavity, lake* lacunae, the spaces occupied by cells of cartilage and bone tissue

lamell- *small plate* concentric lamellae, rings of bone matrix in compact bone

lamina- *layer, sheet* basal lamina, part of the epithelial basement membrane

lat- *wide* latissimus dorsi, a broad muscle of the back

laten- *hidden* latent period of a muscle twitch

later- *side* lateral (directional term)

leuko- *white* leukocyte, white blood cell

leva- *raise, elavate* levator labii superioris, muscle that elevates upper lip

lingua- *tongue* lingual tonsil, adjacent to the tongue

lip-, lipo- *fat, lipid* lipophage, a cell that has taken up fat in its cytoplasm

lith- *stone* cholelithiasis. gallstones

luci- *clear* stratum lucidum, clear layer of the epidermis

lumen- *light* lumen, center of a hollow structure

lut- *yellow* corpus luteum, a yellow hormone-secreting structure in the ovary

lymph- *water* lymphatic circulation, return of clear fluid to the bloodstream

macro- *large* macromolecule, large molecule

macula- *spot* macula lutea, yellow spot on the retina

magn- *large* foramen magnum, largest opening of the skull

mal- *bad, abnormal* malfunction, abnormal functioning of an organ

mamm- *breast* mammary gland, breast

mast- *breast* mastectomy, removal of a mammary gland

mater- *mother* dura mater, pia mater, membranes that envelop the brain

meat- *a passage* external auditory meatus, the ear canal

medi- *middle* medial (directional term)

medull- *marrow* medulla, the middle portion of the kidney, adrenal gland, and lymph node

mega- *large* megakaryocyte, large precursor cell of platelets

meio- *less* meiosis, nuclear division that halves the chromosome number

melan- *black* melanocytes, which secrete the black pigment melanin

men-, menstru- *month* menses, the cyclic menstrual flow

meningo- *membrane* meningitis, inflammation of the membranes of the brain

meso- *middle* mesoderm,. middle germ layer

meta- *beyond, between, transition* metatarsus, the part of the foot between the tarsus and the phalanges

metro- *uterus* metroscope, instrument for examining the uterus

micro- *small* microscope, an instrument used to make small objects appear larger

mictur- *urinate* micturition, the act of voiding the bladder

mito- *thread, filament* mitochondria, small, filament like structures located in cells

mnem- *memory* amnesia

mono- *single* monospasm, spasm of a single limb

morpho- *form* morphology, the study of form and structure of organisms

multi- *many* multinuclear, having several nuclei

mural- *wall* intramural ganglion, a nerve junction within an organ

muta- *change* mutation, change in the base sequence of DNA

myelo- *spinal cord, marrow* myeloblasts, cells of the bone marrow

myo- *muscle* myocardium, heart muscle

nano- *dwarf* nanometer, one billionth of a meter

narco- *numbness* narcotic, a drug producing stupor or numbed sensations

natri- *sodium* atrial natriuretic factor, a sodium-regulating hormone

necro- *death* necrosis tissue death

neo- *new* neoplasm, an abnormal growth

nephro- *kidney* nephritis, inflammation of the kidney

neuro- *nerve* neurophysiology, the physiology of the nervous system

noci- *harmful* nociceptors, receptors for pain

nom- *name* innominate artery; innominate bone

noto- *back* notochord, the embryonic structure that precedes the vertebral column

nucle- *pit, kernel, little nut* nucleus

nutri- *feed, nourish* nutrition

ob- *before, against* obstruction, impeding or blocking up

oculo- *eye* monocular, pertaining to one eye

odonto- *teeth* orthodontist, one who specializes in proper positioning of the teeth in relation to each other

olfact- *smell* olfactory nerves

oligo- *few* oligodendrocytes, neuroglial cells with few branches

onco- *a mass* oncology, study of cancer

oo- *egg* oocyte, precursor of female gamete

ophthalmo- *eye* ophthalmology, the study of the eyes and related disease

orb- *circular* orbicularis oculi, muscle that encircles the eye

orchi- *testis* cryptorchidism, failure of the testes to descend into the scrotum

org- *living* organism

ortho- *straight, direct* orthopedic, correction of deformities of the musculoskeletal system

osm- *smell* anosmia, loss of sense of smell

osmo- *pushing* osmosis

osteo- *bone* osteodermia, bony formations in the skin

oto- *ear* otoscope, a device for examining the ear

ov-, ovi- *egg* ovum, oviduct

oxy- *oxygen* oxygenation, the saturation of a substance with oxygen

pan- *all, universal* panacea, a cure-all

papill- *nipple* dermal papillae, projections of the dermis into the epidermal area

para- *beside, near* paraphrenitis, inflammation of tissues adjacent to the diaphragm

pect-, pectus- *breast* pectoralis major, a large chest muscle

pelv- *a basin* pelvic girdle, which cradles the pelvic organs

peni- *a tail* penis, penile urethra

penna- *a wing* unipennate, bipennate muscles, whose fascicles have a feathered appearance

pent- *five* pentose, a 5-carbon sugar

pep-, peps-, pept- *digest* pepsin, a digestive enzyme of the stomach: peptic ulcer

per-, permea- *through* permeate; permeable

peri- *around* perianal, situated around the anus

phago- *eat* phagocyte, a cell that engulfs and digests particles or cells

pheno- *show, appear* phenotype, the physical appearance of an individual

phleb- *vein* phlebitis, inflammation of the veins

pia- *tender* pia mater, delicate inner membrane around the brain and spinal cord

pili- *hair* arrector pili muscles of the skin, which make the hairs stand erect

pin-, pino- *drink* pinocytosis, the process of a cell in small particles

platy- *flat, broad* platysma, broad, flat muscle of the neck

pleur- *side, rib* pleural serosa, the membrane that lines the thoracic cavity and covers the lungs

plex- plexus- *net, network* brachial plexus the network of nerves that supplies the arm

pneumo- *air, wind* pneumothorax air in the thoracic cavity

pod- *foot* podiatry, the treatment of foot disorders

poly- *multiple* polymorphism multiple forms

post- *after, behind* posterior, places behind (a specific) part

pre-, pro- *before, ahead of* prenatal, before birth

procto- *rectum, anus* proctoscope, an instrument for examining the rectum

pron- *bent forward* prone; pronate

propri- *one's own* proprioception, awareness of body parts and movement

pseudo- *false* pseudotumor, a false tumor

psycho- *mind, psyche* psychogram, a chart of personality traits .

ptos- *fall* renal ptosis, a condition in which the kidneys drift below their normal position

pub- *of the pubis* puberty

pulmo- *lung* pulmonary artery, which brings blood to the lungs

pyo- *pus* pyocyst, a cyst that contains pus

pyro- *fire* pyrogen, a substance that induces fever

quad- quadr- *four-sided* quadratus lumborum. a muscle with a square shape

re- *back, again* reinfect

rect- *straight* rectus abdominis, rectum

ren- *kidney* renal, rennin, an enzyme secreted by the kidney

retin- retic- *net, network* endoplasmic reticulum, a network of membranous sacs within a cell

retro- *backward, behind* retrogression, to move backward in development

rheum- *watery flow, change or flux* rheumatoid arthritis, rheumatic fever

rhin- rhino- *nose* rhinitis, inflammation of the nose

ruga- *fold, wrinkle* rugae. the folds of the stomach, gallbladder and urinary bladder

sagitta- *arrow* sagittal (directional term)

salta- *leap* saltatory conduction, the rapid conduction of impulse along myelinated neurons

sanguine- *blood* consanguineous, indicative of a genetic relationship between individuals

sarco- *flesh* sarocomere, unit of contraction in skeletal muscle

saphen- *visible, clear* great saphenous vein, superficial vein of the thigh and leg

sclero- *hard* sclerodermatitis, inflammitory thickening and hardening of the skin

seb- *grease* sebum, the oil of the skin

semen- *seed, sperm* semen, the discharge of the male reproductive system

semi- *half* semicircular, having the form of half a circle

sens- *feeling* sensation; sensory

septi- *rotten* sepsis, infection; antiseptic

septum- *fence* nasal septum

sero- *serum* serological tests, which assess blood conditions

serrat- *saw* serratus anterior, a muscle of the chest wall that has a jagged edge

sin-, sino- *a hallow* sinuses of the skull

soma- *body* somatic nervous system

somnus- *sleep* insomnia, inability to sleep

sphin- *squeeze* sphincter

splanchn- *organ* splanchnic nerve, autonomic supply to abdominal viscera

spondyl- *vertebra* ankylosing spondylitis, rheumatoid arthritis affecting the spine

squam- *scale, flat* squamous epithelium, squamous suture of skull

steno- *narrow* stenocariasis, narrowing of the pupil

strat- *layer* strata of the epidermis, stratified epithelium

stria- *furrow, streak* striations of skeletal and cardiac muscle tissue

stroma- *spread out* strome, the connective tissue framework of some organs

sub- *beneath, under* sublingual, beneath the tongue

sucr- *sweet* sucrose, table sugar

sudor- *sweat* sudoriferous glands, the sweat glands

super- *above, upon* superior, quality or state of being above others or a part

supra- *above, upon* supracondylar, above a condyle

sym-, syn- *together, with* synapse, the region of communication between two neurons

synerg- *work together* synergism

systol- *contraction* systole, contraction of the heart

tachy- *rapid* tacilycardia, abnormally rapid heartbeat

tact- *touch* tactile sense

telo- *the end* telophase, the end of mitosis

tempi-, tempo- *time* temporal summation of nerve impulses

tens- *stretched* muscle tension

tertius- *third* peroneus tertius, one of three peroneus muscles

tetan- *rigid, tense* tetanus of muscles

therm- *heat* thermometer, an instrument used to measure heat

thromb- *clot* thrombocyte; thrombus

thyro- *a shield* thyroid gland

tissu- *woven* tissue

tono- *tension* tonicity; hypertonic

tax- *poison* antitoxic, effective against poison

trab- *beam, timber* trabeculae, spicules of bone in spongy bone tissue

trans- *across, through* transpleural, through the pleura

trapez- *table* trapezius, the four-sided muscle of the upper back

tri- *three* trifurcation, division into three branches

trop- *turn, change* tropic hormones, whose targets are endocrine glands

troph- *nourish* trophoblast, from which develops the fetal portion of the placenta

tuber- *swelling* tuberosity, a bump on a bone

tunic- *covering* tunica albuginea, the covering of the testis

tympan- *drum* tympanic membrane, the eardrum

ultra- *beyond* ultraviolet radiation, beyond the band of visible light

vacc- *cow* vaccine

vagin- *a sheath* vagina

vagus *wanderer* the vagus nerve, which starts at the brain and travels into the abdominopelvic cavity

valen- *strength* valence shells of atoms

venter- ventr- *hollow cavity, belly* ventral (directional term); ventricle

ventus- *the wind* pulmonary ventilation

vert- *turn* vertical column

vestibul- *a porch* vestibule, the anterior entryway to the mouth and nose

vibr- *shake, quiver* vibrissae, hairs of the nasal vestibule

villus- *sham hair* microvilli, which have the appearance of hair in light microscopy

viscero- *organ, viscera* visceroinhibitory, inhibiting the movements of the viscera

viscos- *sticky* viscosity, resistance to flow

vita- *life* vitamin

vitre- *glass* vitreous humor, the clear jelly of the eye

viv- *live* in vivo

vulv- *a covering* vulva, the female external genitalia

zyg- *a yoke, twin* zygote

SUFFIXES

-able *able to, capable of* viable, ability to live or exist

-ac *referring to* cardiac, referring to the heart

-algia *pain in a certain part* neuralgia, pain along the course of a nerve

-apsi *juncture* synapse, where two neurons connect

-ary *associated with, relating to* coronary, associated with the heart.

-asthen *weakness* myasthenia gravis, a disease involving paralysis

-atomos *indivisible* anatomy, which involves dissection

-bryo *swollen* embryo

-cide *destroy or kill* germicide, an agent that kills germs

-cipit *head* occipital

-clast *break* osteoclast, a cell which dissolves bone matrix

-aine *separate* endocrine organs, which secrete hormones into the blood

-dips *thirst, dry* polydipsia, excessive thirst associated with diabetes

-ectomy *cutting out, surgical removal* appendectomy, cutting out of the appendix

-ell. -elle *small* organelle

-emia *condition of the blood* anemia, deficiency of red blood cells

-esthesi *sensation* anesthesia, lack of sensation

-ferent *carry* efferent nerves, nerves carrying impulses away from the CNS

-form. -forma *shape* cribriform plate of the ethmoid bone

-fuge *driving out* vermifuge. a substance that expels worms of the intestine

-gen *an agent that initiates* pathogen, any agent that produces disease

-glea, -glia *glue* neuroglia, the connective tissue of the nervous system_ **-gram** *data that are systematically recorded, a record* electrocardiogram, a recording showing action of the heart

-sraph *an instrument used for recording data or writing* electrocardiograph, an instrument used to make an electrocardiogram

-ia *condition* insomnia, condition of not being able to sleep

-iatrics *medical specialty* geriatrics, the branch of medicine dealing with disease associated with old age

-ism *condition* hyperthyroidism

-itis *inflammation* gastritis, inflammation of the stomach

-lemma *sheath, husk* sarcolemma, the plasma membrane of a muscle cell

-logy *the study of* pathology, the study of changes in structure and function brought on by disease

-lysis *loosening or breaking down* hydrolysis, chemical decomposition of a compound into other compounds as a result of taking up water

-malacia *soft* osteomalacia, a process leading to bone softening

-mania *obsession, compulsion* erotomania, exaggeration of the sexual passions

-nata *birth* prenatal development

-nom *govern* autonomic nervous system

-odyn *pain* coccygodynia, pain in the region of the coccyx

-oid *like, resembling* cuboid, shaped as a cube

-oma *tumor* lymphoma a tumor of the lymphatic tissues

-opia *defect of the eye* myopia, nearsightedness

-ory *referring to, of* auditory, referring to hearing

-pathy *disease* osteopathy, any disease of the bone

-phasia *speech* aphasia, lack of ability to speak

-phil, -philo *like, love* hydrophilic, water-attracting molecules

-phobia *fear* acrophobia, fear of heights

-phragm *partition* diaphragm, which separates the thoracic and abdominal cavities

-phylax *guard, preserve'* anaphylaxis, prophylactic

-plas *grow* neoplasia, an abnormal growth

-plasm *form, shape* cytoplasm

-plasty *reconstruction of a part, plastic surgery* rhinoplasty, reconstruction of the nose through surgery

-plegia *paralysis* paraplegia, paralysis of the lower half of the body or limbs

-rrhagia *abnormal or excessive discharge* metrorrhagia, uterine hemorrhage

-rrhea *flow or discharge* diarrhea, abnormal emptying of the bowels

-scope *instrument used for examination* stethoscope, instrument used to listen to sounds of parts of the body

-some *body* chromosome

-sorb *suck in* absorb

-stalsis *compression* peristalsis, muscular contractions that propel food along the digestive tract

-stasis *arrest, fixation* hemostasis, arrest of bleeding

-stitia *come to stand* interstitial fluid, between the cells

-stomy *establishment of an artificial opening* enterostomy, the formation of an artificial opening into the intestine through the abdominal wall

-tegm *cover* integument

-tomy *to cut* appendectomy, surgical removal of the appendix

-trud *thrust* protrude, detrusor muscle

-ty *condition of, state* immunity, condition of being resistant to infection or disease

-uria *urine* polyuria, passage of an excessive amount of urine

-zyme *ferment* enzyme

Appendix: Word Roots, Prefixes, Suffixes, and Combining Forms

PREFIXES AND COMBINING FORMS

a-, an- *absence or lack* acardia, lack of a heart: anaerobic, in the absence of oxygen

ab- *departing from: away from* abnormal, departing from normal

acou- *hearing* acoustics, the science of sound

acr-,acro- *extreme or extremity; peak* acrodermatitis, inflammation of the skin of the extremities

ad- *to or toward* adorbital, toward the orbit

aden-, adeno- *gland* adeniform, resembling a gland in shape

adren- *toward the kidney* adrenal gland, adjacent to the kidney

aero- *air* aerobic respiration; oxygen-requiring metabolism

af- *toward* afferent neurons, which carry impulses to the central nervous system

agon- *contest* agonistic and antagonistic muscles, which oppose each other

alb- *white* corpus albicans of the ovary, a white scar tissue

aliment- *nourish* alimentary canal or digestive tract

allel- *of one another* alleles, alternative expressions of a gene

amphi- *on both sides; of both kinds* amphibian, an organism capable of living in water and on land

ana- *apart, up; again* anaphase of mitosis, when the chromosomes separate

anastomos- *come together* arteriovenous anastomosis, a connection between an artery and a vein

aneurysm *a widening* aortic aneurism, a weak spot that causes enlargement of the blood vessel

angi- *vessel* angiitis. inflammation of a lymph vessel or blood vessel

angin- *choked* angina pectoris, a choked feeling in the chest due to dysfunction of the heart

ant-, anti- *opposed to; preventing or inhibiting* anticoagulant, a substance that prevents blood coagulation

ante- *preceding, before* antecubital, in front of the elbow

aort- *great artery* aorta

ap-, api- *tip, extremity* apex of the heart

append- *hang to* appendicular skeleton

aqua-, aque- *water* aqueous solutions

arbor- *tree* arbor vitae of the cerebellum, the tree-like pattern of white matter

areola- *open space* areolar connective tissue, a loose connective tissue.

arrect- *upright* arrector pili muscles of the skin, which make the hairs stand erect

arthr-, arthro- *joint* arthropathy, any joint disease

artic-. *joint* articular surfaces of bones, the points of connection

atri- *vestibule* atria, upper chambers of the heart

auscult- *listen* auscultatory method for measuring blood pressure

aut- auto- *self* autogenous, self-generated

ax- axi- axo- *axis, axle* axial skeleton, axis of vertebral column

azyg- *unpaired* azygous vein, an unpaired vessel

baro- *pressure* baroreceptors for monitoring blood pressure

basal- *base* basal lamina of epithelial basement membrane

bi- *two* bicuspid, having two cusps

bili- *bile* bilirubin, a bile pigment

bio- *life* biology, the study of life and living organisms

blast- *bud or germ* blastocyte, undifferentiated embryonic cell

brachi- *arm* brachial plexus of peripheral nervous system supplies the arm

brady- *slow* bradycardia, abnormally slow heart rate

brev- *short* peroneus brevis, a short leg muscle

broncha- *bronchus* bronchospasm, spasmodic contraction of bronchial muscle

bucco- *cheek* buccolabial, pertaining to the cheek and lip

calor- *heat* calories, a measure of energy

capill- *hair* blood and lymph capillaries

caput- *head* decapitate, remove the head

carcin- *cancer* carcinogen, a cancer-causing agent

cardi, cardio- *heart* cardiotoxic, harmful to the heart

carneo- *flesh* trabeculae carneae, ridges of muscle in the ventricles of the heart

carat 1) *carrot,* 2) *stupor* carotene, an orange pigment; 2) carotid arteries in the neck, blockage causes fainting

cata- *down* catabolism, chemical breakdown

caud- *tail* caudal (directional term)

cec- *blind* cecum of large intestine, a blind-ended pouch

cele- *abdominal* celiac artery, in the abdomen

cephal- *head* cephalometer, an instrument for measuring the head

cerebro- *brain, especially the cerebrum* cerebrospinal, pertaining to the brain and spinal cord

cervic-, cervix- *neck* cervix of the uterus

chiasm- *crossing* optic chiasma, where optic nerves cross

chole- *bile* cholesterol; cholecystokinin, a bile-secreting hormone

chondr- *cartilage* chondrogenic, giving rise to cartilage

chrom- *colored* chromosome, so named because they stain darkly

cili- *small hair* ciliated epithelium

circum- *around* circumnuclear, surrounding the nucleus

clavic- *key* clavicle, a "skeleton key"

co-, con- *together* concentric, common center, together in the center

coccy- *cuckoo* coccyx, which is beak-shaped

cochlea- *snail shell* the cochlea of the inner ear, which is coiled like a snail shell

coel- *hallow* coelom, the ventral body cavity

commis- *united* gray commissure of the spinal cord connects the two columns of gray matter

concha- *shell* nasal conchae, coiled shelves of bone in the nasal cavity

contra- *against* contraceptive, agent preventing conception

corn-, cornu- *horn* stratum corneum, outer layer of the skin composed of (horny) cells

corona- *crown* coronal suture of the skull

corp- *body* corpse. corpus luteum, hormone-secreting body in the ovary

cort- *bark* cortex, the outer layer of the brain, kidney, adrenal glands, lymph nodes, ovaries

cost- *rib* intercostal, between the ribs

crani- *skull* craniotomy, a skull operation

crypt- *hidden* cyptomenorrhea, a condition in which menstrual symptoms are experienced but no external loss of blood occurs

cusp- *pointed* bicuspid, tricuspid valves of the heart

cutic- *skin* cuticle of the nail

cyan- *blue* cyanosis, blue color of the skin due to lack of oxygen

cyst- *sac, bladder* cystitis, inflammation of the urinary bladder

cyt- *cell* cytology, the study of cells

de- *undoing. reversal, loss, removal* deactivation, becoming inactive

decid- *falling off* deciduous (milk) teeth

delta- *triangular* deltoid muscle, roughly triangular in shape

den-, dent- *tooth* dentin of the tooth

dendr- *tree, branch* dendrites, telodendria, both branches of a neuron

derm- *skin* dermis, deep layer of the skin

desm- *bond* desmosome, which binds adjacent epithelial cells

di- *twice, double* dimorphism, having two forms

dia- *through, between* diaphragm, the wall through or between two areas

dialys- *separate, break apart* kidney dialysis, in which waste products are removed from the blood

diastol- *stand apart* cardiac diastole, between successive contractions of the heart

diure- *urinate* diuretic, a drug that increases urine output

dors- *the back* dorsal; dorsum; dorsiflexion

duc-, duct- *lead, draw* ductus deferens which carries sperm from the epididymus into the urethra during ejaculation

dura- *hard* dura mater, tough outer meninx

dys- *difficult, faulty, painful* dyspepsia, disturbed digestion

ec-, ex-, ecto- *out, outside, awayfrom* excrete, to remove materials from the body

ectop- *displaced* ectopic pregnancy, ectopic focus for initiation of heart contraction

edem- *swelling* edema, accumulation of water in body tissues

ef- *away* efferent nerve fibers, which carry impulses away from the central nervous system

ejac- *to shoot forth* ejaculation of semen

embol- *wedge* embolus, an obstructive object traveling in the bloodstream

en-, em- *in, inside* encysted, enclosed in a cyst or capsule

enceph.- *brain* encephalitis, inflammation of the brain

endo- *within, inner* endocytosis, taking particles into a cell

entero- *intestine* enterologist, one who specializes in the study of intestinal disorders

epi- *over, above* epidermis, outer layer of skin

erythr- *red* erythema, redness of the skin; erythrocyte, red blood cell

eso- *within* esophagus

eu- *well* euesthesia, a normal state of the senses

excret- *separate* excretory system

exo- *outside, outer layer* exophthalmos, an abnormal protrusion of the eye from the orbit

extra- *outside, beyond* extracellular, outside the body cells of an organism.

extrins- *from the outside* extrinsic regulation of the heart

fasci-, fascia- *bundle, band* superficial and deep fascia

fenestr- *window* fenestrae of the inner ear; fenestrated capillaries

ferr- *iron* transferrin, ferritin, both iron-storage proteins

flagell- *whip* flagellum, the tail of a sperm cell

flat- *blow, blown* flatulence

folli- *bag, bellows* hair follicle

fontan- *fountain* fontanels of the fetal skull

foram- *opening-* foramen magnum of the skull

foss- *ditch* fossa ovalis of the heart: mandibular fossa of the skull

gam- gamet- *married, spouse* gametes, the sex cells

gangli- *swelling, or knot* dorsal root ganglia of the spinal nerves

gastr- *stomach* gastrin, a hormone that influences gastric acid secretion

gene- *beginning; origin* genetics

germin- *grow* germinal epithelium of the gonads

gem-, geront- *old man* gerontology, the study of aging

gest- *carried* gestation, the period from conception to birth

glauc- *gray* glaucoma, which causes gradual blindness

glom- *ball* glomeruli, clusters of capillaries in the kidneys

glosso- *tongue* glossopathy, any disease of the tongue

gluco-, glyco- gluconeogenesis. the production of glucose from noncarbohydrate molecules

glute- *buttock* gluteus maximus. the largest muscle of the buttock

gnost- *knowing* the gnostic sense, a sense of awareness of self

gompho- *nail* gomphosis, the term applied to the joint between tooth and jaw

gon- gono- *seed, offspring* gonads, the sex organs

gust- *taste* gustatory sense, the sense of taste

hapt- *fasten, grasp* hapten, a partial antigen

hema-, hemato., hemo- *blood* hematocyst. a cyst containing blood

hemi- *half* hemiglossal, pertaining to one-half of the tongue

hepat- *liver* hepatitis, inflammation of the liver

hetero- *different or other* heterosexuality, sexual desire for a person of the opposite sex

hiat *gap* the hiatus of the diaphragm, the opening through which the esophagus passes

hippo- *horse* hippocampus of the brain, shaped like a seahorse .

hirsut- *hairy* hirsutism, excessive body hair

hist- *tissue* histology, the study of tissues

holo- *whole* holocrine glands, whose secretions are whole cells

hom-, homo- *same* homeoplasia, formation of tissue similar to normal tissue; homocentric, having the same center

hormon- *to excite* hormones

humor- *a fluid* humoral immunity, which involves antibodies circulating in the blood

hyal- *clear* hyaline cartilage, which has no visible fibers

hydr-, hydro- *water* dehydration, loss of body water

hyper- *excess* hypertension, excessive tension

hypno- *sleep* hypnosis, a sleeplike state

hypo- *below, deficient* hypodermic, beneath the skin; hypokalemia, deficiency of potassium

hyster-, hystero- *uterus or womb* hysterectomy, removal of the uterus; hysterodynia. pain in the womb

ile- *intestine* ileum, the last portion of the small intestine

im- *not* impermeable, not permitting passage, not permeable

inter- *between* intercellular, between the cells

intercal- *insert* intercalated disc, the end membranes between adjacent cardiac muscle

intra- *within, inside* intracellular, inside the cell

iso- *equal, same* isothermal, equal or same temperature

jugul- *throat* jugular veins, prominent vessels in the neck

juxta- *near, close to* juxtaglomerular apparatus, a cell cluster next to the glomeruli in the kidneys

karyo- *kernal nucleus* karyotype, the assemblage of the nuclear chromosomes

kera- *horn* keratin, the water-repellent protein of the skin

kilo- *thousand* kilocalories, equivalent to one thousand calories

kin- kines- *move* kinetic energy, the energy of motion

lahi- lahri- *lip* labial frenulum. the membrane which joins the lip to the gum

lact- *milk* lactose milk sugar

lacun- *space, cavity, lake* lacunae, the spaces occupied by cells of cartilage and bone tissue

lamell- *small plate* concentric lamellae, rings of bone matrix in compact bone

lamina- *layer, sheet* basal lamina, part of the epithelial basement membrane

lat- *wide* latissimus dorsi, a broad muscle of the back

laten- *hidden* latent period of a muscle twitch

later- *side* lateral (directional term)

leuko- *white* leukocyte, white blood cell

leva- *raise, elavate* levator labii superioris, muscle that elevates upper lip

lingua- *tongue* lingual tonsil, adjacent to the tongue

lip-, lipo- *fat, lipid* lipophage, a cell that has taken up fat in its cytoplasm

lith- *stone* cholelithiasis. gallstones

luci- *clear* stratum lucidum, clear layer of the epidermis

lumen- *light* lumen, center of a hollow structure

lut- *yellow* corpus luteum, a yellow hormone-secreting structure in the ovary

lymph- *water* lymphatic circulation, return of clear fluid to the bloodstream

macro- *large* macromolecule, large molecule

macula- *spot* macula lutea, yellow spot on the retina

magn- *large* foramen magnum, largest opening of the skull

mal- *bad, abnormal* malfunction, abnormal functioning of an organ

mamm- *breast* mammary gland, breast

mast- *breast* mastectomy, removal of a mammary gland

mater- *mother* dura mater, pia mater, membranes that envelop the brain

meat- *a passage* external auditory meatus, the ear canal

medi- *middle* medial (directional term)

medull- *marrow* medulla, the middle portion of the kidney, adrenal gland, and lymph node

mega- *large* megakaryocyte, large precursor cell of platelets

meio- *less* meiosis, nuclear division that halves the chromosome number

melan- *black* melanocytes, which secrete the black pigment melanin

men-, menstru- *month* menses, the cyclic menstrual flow

meningo- *membrane* meningitis, inflammation of the membranes of the brain

meso- *middle* mesoderm,. middle germ layer

meta- *beyond, between, transition* metatarsus, the part of the foot between the tarsus and the phalanges

metro- *uterus* metroscope, instrument for examining the uterus

micro- *small* microscope, an instrument used to make small objects appear larger

mictur- *urinate* micturition, the act of voiding the bladder

mito- *thread, filament* mitochondria, small, filament like structures located in cells

mnem- *memory* amnesia

mono- *single* monospasm, spasm of a single limb

morpho- *form* morphology, the study of form and structure of organisms

multi- *many* multinuclear, having several nuclei

mural- *wall* intramural ganglion, a nerve junction within an organ

---

muta- *change* mutation, change in the base sequence of DNA

myelo- *spinal cord, marrow* myeloblasts, cells of the bone marrow

myo- *muscle* myocardium, heart muscle

nano- *dwarf* nanometer, one billionth of a meter

narco- *numbness* narcotic, a drug producing stupor or numbed sensations

natri- *sodium* atrial natriuretic factor, a sodium-regulating hormone

necro- *death* necrosis tissue death

neo- *new* neoplasm, an abnormal growth

nephro- *kidney* nephritis, inflammation of the kidney

neuro- *nerve* neurophysiology, the physiology of the nervous system

noci- *harmful* nociceptors, receptors for pain

nom- *name* innominate artery; innominate bone

noto- *back* notochord, the embryonic structure that precedes the vertebral column

nucle- *pit, kernel, little nut* nucleus

nutri- *feed, nourish* nutrition

ob- *before, against* obstruction, impeding or blocking up

oculo- *eye* monocular, pertaining to one eye

odonto- *teeth* orthodontist, one who specializes in proper positioning of the teeth in relation to each other

olfact- *smell* olfactory nerves

oligo- *few* oligodendrocytes, neuroglial cells with few branches

onco- *a mass* oncology, study of cancer

oo- *egg* oocyte, precursor of female gamete

ophthalmo- *eye* ophthalmology, the study of the eyes and related disease

orb- *circular* orbicularis oculi, muscle that encircles the eye

orchi- *testis* cryptorchidism, failure of the testes to descend into the scrotum

org- *living* organism

ortho- *straight, direct* orthopedic, correction of deformities of the musculoskeletal system

osm- *smell* anosmia, loss of sense of smell

osmo- *pushing* osmosis

osteo- *bone* osteodermia, bony formations in the skin

oto- *ear* otoscope, a device for examining the ear

ov-, ovi- *egg* ovum, oviduct

oxy- *oxygen* oxygenation, the saturation of a substance with oxygen

pan- *all, universal* panacea, a cure-all

papill- *nipple* dermal papillae, projections of the dermis into the epidermal area

para- *beside, near* paraphrenitis, inflammation of tissues adjacent to the diaphragm

pect-, pectus- *breast* pectoralis major, a large chest muscle

pelv- *a basin* pelvic girdle, which cradles the pelvic organs

peni- *a tail* penis, penile urethra

penna- *a wing* unipennate, bipennate muscles, whose fascicles have a feathered appearance

pent- *five* pentose, a 5-carbon sugar

pep-, peps-, pept- *digest* pepsin, a digestive enzyme of the stomach: peptic ulcer

per-, permea- *through* permeate; permeable

peri- *around* perianal, situated around the anus

phago- *eat* phagocyte, a cell that engulfs and digests particles or cells

pheno- *show, appear* phenotype, the physical appearance of an individual

phleb- *vein* phlebitis, inflammation of the veins

pia- *tender* pia mater, delicate inner membrane around the brain and spinal cord

pili- *hair* arrector pili muscles of the skin, which make the hairs stand erect

pin-, pino- *drink* pinocytosis, the process of a cell in small particles

platy- *flat, broad* platysma, broad, flat muscle of the neck

pleur- *side, rib* pleural serosa, the membrane that lines the thoracic cavity and covers the lungs

plex- plexus- *net, network* brachial plexus the network of nerves that supplies the arm

pneumo- *air, wind* pneumothorax air in the thoracic cavity

pod- *foot* podiatry, the treatment of foot disorders

poly- *multiple* polymorphism multiple forms

post- *after, behind* posterior, places behind (a specific) part

pre-, pro- *before, ahead of* prenatal, before birth

procto- *rectum, anus* proctoscope, an instrument for examining the rectum

pron- *bent forward* prone; pronate

propri- *one's own* proprioception, awareness of body parts and movement

pseudo- *false* pseudotumor, a false tumor

psycho- *mind, psyche* psychogram, a chart of personality traits .

ptos- *fall* renal ptosis, a condition in which the kidneys drift below their normal position

pub- *of the pubis* puberty

pulmo- *lung* pulmonary artery, which brings blood to the lungs

pyo- *pus* pyocyst, a cyst that contains pus

pyro- *fire* pyrogen, a substance that induces fever

quad- quadr- *four-sided* quadratus lumborum. a muscle with a square shape

re- *back, again* reinfect

rect- *straight* rectus abdominis, rectum

ren- *kidney* renal, rennin, an enzyme secreted by the kidney

retin- retic- *net, network* endoplasmic reticulum, a network of membranous sacs within a cell

retro- *backward, behind* retrogression, to move backward in development

rheum- *watery flow, change or flux* rheumatoid arthritis, rheumatic fever

rhin- rhino- *nose* rhinitis, inflammation of the nose

ruga- *fold, wrinkle* rugae. the folds of the stomach, gallbladder and urinary bladder

sagitta- *arrow* sagittal (directional term)

salta- *leap* saltatory conduction, the rapid conduction of impulse along myelinated neurons

sanguine- *blood* consanguineous, indicative of a genetic relationship between individuals

sarco- *flesh* sarocomere, unit of contraction in skeletal muscle

saphen- *visible, clear* great saphenous vein, superficial vein of the thigh and leg

sclero- *hard* sclerodermatitis, inflammitory thickening and hardening of the skin

seb- *grease* sebum, the oil of the skin

semen- *seed, sperm* semen, the discharge of the male reproductive system

semi- *half* semicircular, having the form of half a circle

sens- *feeling* sensation; sensory

septi- *rotten* sepsis, infection; antiseptic

septum- *fence* nasal septum

sero- *serum* serological tests, which assess blood conditions

serrat- *saw* serratus anterior, a muscle of the chest wall that has a jagged edge

sin-, sino- *a hallow* sinuses of the skull

soma- *body* somatic nervous system

somnus- *sleep* insomnia, inability to sleep

sphin- *squeeze* sphincter

splanchn- *organ* splanchnic nerve, autonomic supply to abdominal viscera

spondyl- *vertebra* ankylosing spondylitis, rheumatoid arthritis affecting the spine

squam- *scale, flat* squamous epithelium, squamous suture of skull

steno- *narrow* stenocariasis, narrowing of the pupil

strat- *layer* strata of the epidermis, stratified epithelium

stria- *furrow, streak* striations of skeletal and cardiac muscle tissue

stroma- *spread out* strome, the connective tissue framework of some organs

sub- *beneath, under* sublingual, beneath the tongue

sucr- *sweet* sucrose, table sugar

sudor- *sweat* sudoriferous glands, the sweat glands

super- *above, upon* superior, quality or state of being above others or a part

supra- *above, upon* supracondylar, above a condyle

sym-, syn- *together, with* synapse, the region of communication between two neurons

synerg- *work together* synergism

systol- *contraction* systole, contraction of the heart

tachy- *rapid* tacilycardia, abnormally rapid heartbeat

tact- *touch* tactile sense

telo- *the end* telophase, the end of mitosis

tempi-, tempo- *time* temporal summation of nerve impulses

tens- *stretched* muscle tension

tertius- *third* peroneus tertius, one of three peroneus muscles

tetan- *rigid, tense* tetanus of muscles

therm- *heat* thermometer, an instrument used to measure heat

thromb- *clot* thrombocyte; thrombus

thyro- *a shield* thyroid gland

tissu- *woven* tissue

tono- *tension* tonicity; hypertonic

tax- *poison* antitoxic, effective against poison

trab- *beam, timber* trabeculae, spicules of bone in spongy bone tissue

trans- *across, through* transpleural, through the pleura

trapez- *table* trapezius, the four-sided muscle of the upper back

tri- *three* trifurcation, division into three branches

trop- *turn, change* tropic hormones, whose targets are endocrine glands

troph- *nourish* trophoblast, from which develops the fetal portion of the placenta

tuber- *swelling* tuberosity, a bump on a bone

tunic- *covering* tunica albuginea, the covering of the testis

tympan- *drum* tympanic membrane, the eardrum

ultra- *beyond* ultraviolet radiation, beyond the band of visible light

vacc- *cow* vaccine

vagin- *a sheath* vagina

vagus *wanderer* the vagus nerve, which starts at the brain and travels into the abdominopelvic cavity

valen- *strength* valence shells of atoms

venter- ventr- *hollow cavity, belly* ventral (directional term); ventricle

ventus- *the wind* pulmonary ventilation

vert- *turn* vertical column

vestibul- *a porch* vestibule, the anterior entryway to the mouth and nose

vibr- *shake, quiver* vibrissae, hairs of the nasal vestibule

villus- *sham hair* microvilli, which have the appearance of hair in light microscopy

viscero- *organ, viscera* visceroinhibitory, inhibiting the movements of the viscera

viscos- *sticky* viscosity, resistance to flow

vita- *life* vitamin

vitre- *glass* vitreous humor, the clear jelly of the eye

viv- *live* in vivo

vulv- *a covering* vulva, the female external genitalia

zyg- *a yoke, twin* zygote

SUFFIXES

-able *able to, capable of* viable, ability to live or exist

-ac *referring to* cardiac, referring to the heart

-algia *pain in a certain part* neuralgia, pain along the course of a nerve

-apsi *juncture* synapse, where two neurons connect

-ary *associated with, relating to* coronary, associated with the heart.

-asthen *weakness* myasthenia gravis, a disease involving paralysis

-atomos *indivisible* anatomy, which involves dissection

-bryo *swollen* embryo

-cide *destroy or kill* germicide, an agent that kills germs

-cipit *head* occipital

-clast *break* osteoclast, a cell which dissolves bone matrix

-aine *separate* endocrine organs, which secrete hormones into the blood

-dips *thirst, dry* polydipsia, excessive thirst associated with diabetes

-ectomy *cutting out, surgical removal* appendectomy, cutting out of the appendix

-ell. -elle *small* organelle

-emia *condition of the blood* anemia, deficiency of red blood cells

-esthesi *sensation* anesthesia, lack of sensation

-ferent *carry* efferent nerves, nerves carrying impulses away from the CNS

-form. -forma *shape* cribriform plate of the ethmoid bone

-fuge *driving out* vermifuge. a substance that expels worms of the intestine

-gen *an agent that initiates* pathogen, any agent that produces disease

-glea, -glia *glue* neuroglia, the connective tissue of the nervous system -gram *data that are systematically recorded, a record* electrocardiogram, a recording showing action of the heart

-sraph *an instrument used for recording data or writing* electrocardiograph, an instrument used to make an electrocardiogram

-ia *condition* insomnia, condition of not being able to sleep

-iatrics *medical specialty* geriatrics, the branch of medicine dealing with disease associated with old age

-ism *condition* hyperthyroidism

-itis *inflammation* gastritis, inflammation of the stomach

-lemma *sheath, husk* sarcolemma, the plasma membrane of a muscle cell

-logy *the study of* pathology, the study of changes in structure and function brought on by disease

-lysis *loosening or breaking down* hydrolysis, chemical decomposition of a compound into other compounds as a result of taking up water

-malacia *soft* osteomalacia, a process leading to bone softening

-mania *obsession, compulsion* erotomania, exaggeration of the sexual passions

-nata *birth* prenatal development

-nom *govern* autonomic nervous system

-odyn *pain* coccygodynia, pain in the region of the coccyx

-oid *like, resembling* cuboid, shaped as a cube

-oma *tumor* lymphoma a tumor of the lymphatic tissues

-opia *defect of the eye* myopia, nearsightedness

-ory *referring to, of* auditory, referring to hearing

-pathy *disease* osteopathy, any disease of the bone

-phasia *speech* aphasia, lack of ability to speak

-phil, -philo *like, love* hydrophilic, water-attracting molecules

-phobia *fear* acrophobia, fear of heights

-phragm *partition* diaphragm, which separates the thoracic and abdominal cavities

-phylax *guard, preserve'* anaphylaxis, prophylactic

-plas *grow* neoplasia, an abnormal growth

-plasm *form, shape* cytoplasm

-plasty *reconstruction of a part, plastic surgery* rhinoplasty, reconstruction of the nose through surgery

-plegia *paralysis* paraplegia, paralysis of the lower half of the body or limbs

-rrhagia *abnormal or excessive discharge* metrorrhagia, uterine hemorrhage

-rrhea *flow or discharge* diarrhea, abnormal emptying of the bowels

-scope *instrument used for examination* stethoscope, instrument used to listen to sounds of parts of the body

-some *body* chromosome

-sorb *suck in* absorb

-stalsis *compression* peristalsis, muscular contractions that propel food along the digestive tract

-stasis *arrest, fixation* hemostasis, arrest of bleeding

-stitia *come to stand* interstitial fluid, between the cells

-stomy *establishment of an artificial opening* enterostomy, the formation of an artificial opening into the intestine through the abdominal wall

-tegm *cover* integument

-tomy *to cut* appendectomy, surgical removal of the appendix

-trud *thrust* protrude, detrusor muscle

-ty *condition of, state* immunity, condition of being resistant to infection or disease

-uria *urine* polyuria, passage of an excessive amount of urine

-zyme *ferment* enzyme

Made in the USA
Las Vegas, NV
19 August 2022

53530156R00072